Improving Water Governance in Kathmandu: Insights from Systems Thinking and Behavioral Science

In Focus – Special Book Series

Improving Water Governance in Kathmandu: Insights from Systems Thinking and Behavioral Science

Editors
Dale Whittington and Xun Wu

IWA PUBLISHING

Published by **IWA Publishing**
Alliance House
12 Caxton Street
London SW1H 0QS, UK
Telephone: +44 (0)20 7654 5500
Fax: +44 (0)20 7654 5555
Email: publications@iwap.co.uk
Web: www.iwapublishing.com

First published 2020

British Library Cataloguing in Publication Data
A CIP catalogue record for this book is available from the British Library

ISBN: 9781789061451 (paperback)
ISBN: 9781789061468 (eBook)
ISBN: 9781789061475 (ePUB)

Contents

Editorial
Improving water governance in Kathmandu: insights from systems thinking and behavioral science

Dale Whittington[a] and Xun Wu[b]

[a]*Corresponding author. Departments of Environmental Sciences & Engineering and City & Regional Planning, University of North Carolina at Chapel Hill, Chapel Hill, NC, USA; Global Development Institute, University of Manchester, Manchester, UK; and Institute of Water Policy, Lee Kwan Yew School of Public Policy, National University of Singapore, Singapore*
E-mail: Dale_Whittington@unc.edu
[b]*Division of Public Policy, Hong Kong University of Science and Technology, Hong Kong*

Introduction

At the time this introduction to the special issue is being written (September 2019), the global water and sanitation community is wrestling with the policy implications of two important realizations. The first is that it is quite possible for cities to actually run out of water – for the piped network to run dry. In the spring of 2018, the world watched the unfolding water crisis in Cape Town, South Africa. The Cape Town water utility's website showed a countdown to day zero when the pipes would run dry in a modern global city of 3.7 million people. Cape Town was just weeks away from this calamity when heavy rains arrived to raise the reservoir level in the city's water supply system. While the rains offered a reprieve, the world had been given a glimpse of the challenges of this emerging era of water scarcity (Muller, 2017). Rapid population and economic growth, coupled with rural to urban migration and climate change, is increasing the threat of water scarcity in many cities in the Global South. Cape Town barely avoided having its piped network run dry, but this near miss should not make us overconfident in our ability to manage water scarcity.

The second realization is that in many locations, basic water and sanitation interventions do not result in the large public health improvements that many water and sanitation professionals had hoped (Arnold *et al.*, 2018; Coffey & Spears, 2018; Cumming & Curtis, 2018; Brown *et al.*, 2019; Pickering *et al.*, 2019; Whittington *et al.*, 2020). The evidence comes from a series of recent randomized controlled trials that evaluated water and sanitation interventions in the Global South[1]. However, we do know

[1] These new findings are, in fact, consistent with much older evidence from observational studies (Esrey, 1996)

doi: 10.2166/wp.2019.000

the water and sanitation intervention that will result in dramatic public health improvements: modern housing connected to a piped water distribution network with 24/7 supply of potable water and piped sewerage collection (Cutler & Miller, 2005). What is not fully appreciated by water and sanitation professionals is how expensive modern housing and piped water and sewer collection actually is.

Households living in modern housing demand four main water-related services: (1) potable water for drinking, cooking, and dish washing; (2) water for bathing; (3) water for clothes washing; and (4) water for feces and urine disposal. In many locations, they would also like water for outdoor gardening or lawn watering. The costs of plumbing and appliances embedded in the cost of modern housing are roughly twice the cost of the public infrastructure needed to supply potable water and carry away and treat the wastewater[2]. The magnitude of these investments required inside modern housing units in order to carry out these four activities is embedded in the cost of the housing and is often invisible to water and sanitation professionals.

However, many low- and middle-income households in cities in the Global South, such as Kathmandu, simply cannot afford these embedded costs of using water in their homes even if they have access to heavily subsidized piped water and sewer services. Some have housing units with indoor kitchens only, with pit latrines outdoors. Others have kitchens and showers, but continue to defecate outside the housing unit. Some housing units have kitchens, showers, and toilet indoors, but wash clothes at facilities outside the home. As incomes grow, households face a complex set of tradeoffs between spending their limited budget on water and sanitation services inside and outside the house and between improved water and sanitation facilities and other attributes of their housing unit (as well as other budget priorities).

Similarly, in cities like Kathmandu, water utilities face complex choices with respect to what kinds of water and sewer services should be provided to houses in neighborhoods with different types of housing. There is little point in providing sewer networks to neighborhoods where households do not have piped water inside their housing units, although this does happen. Similarly, making large investments in public toilets and public taps is likely ill-advised in neighborhoods where households are rapidly investing in modern housing.

As water and sanitation professionals work out the implications of these two realizations on policy and planning for water and sanitation improvements in the Global South, they will require an in-depth knowledge of local housing, water, and sanitation conditions, as well as a nuanced understanding of how households prioritize improvements in housing, water, and sanitation. The papers in this special issue about Kathmandu illustrate the types of analyses of local conditions that are needed. Like Cape Town, Kathmandu holds many lessons for the global community about households' responses to water scarcity and the management of water and sanitation services in periods of rapid urbanization and climate change.

---

[2] For purposes of illustration, if the costs per cubic meter of water and wastewater services are US$5 per cubic meter and a household uses 15 cubic meters per month, the cost per household for the public water and sanitation infrastructure would be US$75 per month. If the costs of water embedded in housing are double the public infrastructure costs, they would be about US$150 per month, resulting in a total cost of using water in a modern housing unit of US$225 per month for these four water-related activities in the home or apartment: (1) potable water for drinking, cooking, and dish washing; (2) water for feces and urine disposal; (3) water for bathing; and (4) water for clothes washing. The costs of public water-related infrastructure are typically heavily subsidized, so households rarely pay the full costs of US$75 per month for the public infrastructure. Subsidies for modern housing, however, are more limited. Many households do pay the full costs of US$150 per month for the plumbing and appliances needed to carry out their desired water-related activities inside the home.

Background: an overview of water and sanitation conditions in Kathmandu

Over the past three decades, Kathmandu has experienced rapid population growth due to massive in-migration that resulted in part from political turmoil and civil war. However, there has been little investment in Kathmandu's municipal water supply system inside the Kathmandu Valley. Thus, the gap between the water supply needs of the population and the ability of the municipal water supply system to meet these needs has been growing. In 2014, about 90% of households interviewed in one study in Kathmandu had a private connection to the public water supply system, but most received little water (Gurung *et al.*, 2017).

Planning for a major infrastructure project to bring new water supplies to Kathmandu – the Melamchi Water Supply Project (MWSP) – started in the 1980s. The project involves an interbasin water transfer of about 170,000 cubic meters per day to be delivered through a 26-km long tunnel through a mountain. The construction of the Melamchi tunnel did not begin until 2009. The project experienced further delays due to problems with multiple foreign contractors. In 2019, residents of Kathmandu have still not received any additional water supplies from outside the Kathmandu Valley. Hopefully, new water supplies will reach Kathmandu soon and alleviate many of Kathmandu's water supply problems (Asian Development Bank, 2013, 2015).

Conditions in Kathmandu offer a unique window on what happens when investments in the piped water supply infrastructure in a large, rapidly growing city in a low-income country are delayed for decades. Households in Kathmandu have responded in two main ways to the increasing urban water scarcity. First, many households have drilled private wells to supplement their water supply from the Kathmandu municipal water utility's (Kathmandu Upatyaka Khanepani Limited – KUKL) piped network and, as a result, groundwater levels are falling in the Kathmandu Valley. Second, households began purchasing increasing quantities of water from a new, rapidly growing private water vending industry. In effect, the private sector investment filled part of the water shortfall that resulted from delays in the public investment in the municipal water supply system.

In 2014, 85% of households in Kathmandu were using two or more water sources. Approximately 30% of household water is supplied from private connections, 29% from private wells, 20% from tankers, 8% from public wells, 6% from public taps, and the rest (7%) from neighbors, bottled water (20-l jars), stone taps, and surface water. On average, households self-supplied or purchased about 85 liters per capita per day (lpcd) in the rainy season and 69 lpcd in the dry season (Gurung *et al.*, 2017). However, such coping strategies have entailed high costs to households (Whittington *et al.*, 2002; Pattanayak *et al.*, 2005). Average household coping costs more than doubled in real terms over the period from 2001 to 2014, from US$5 to US$12 per month, measured in 2014 prices (Gurung *et al.*, 2017), as households struggled to find individual solutions to deal with the problems of Kathmandu's poor quality, unreliable piped water supply system.

Challenges in water governance for water supply augmentation projects

There are high hopes that the MWSP will be a savior for the chronic water source management problems currently facing Kathmandu. However, it should not be assumed that the benefits of the project will be easy to achieve or that they are inevitable. The effectiveness of water supply augmentation projects, such as the MWSP, hinges critically upon the policy framework, institutional arrangements, and management practices (Saleth & Dinar, 1997; Grafton *et al.*, 2015). Water scarcity in many cities in developing countries, such as Kathmandu, is often deeply rooted in deficiencies and misalignments in water governance. Therefore, unless such deficiencies and misalignments are tackled head on, the benefits of water supply augmentation projects, such as the MWSP, may be much smaller than anticipated.

There are several potential routes through which this could happen. First, while water supply augmentation projects increase the amount of water available in the system, the efficiency of the water distribution system will play a key role in determining the size of the ultimate gain from the increased water supply. In many cities in developing countries, the increase in water supply may not be converted to an improvement in the financial situation of the water utility if deficiencies in water tariff systems remain untackled. As a result, limited or no investment may be made to improve the efficiency of the water distribution system by reducing non-revenue water (Sahin *et al.*, 2016).

Second, water scarcity in many cities in developing countries is often accompanied with a high level of inequity in water allocation among households with different social–economic characteristics (Bakker *et al.*, 2008). The existing patterns in water supply provision are dictated by political, institutional, and geographical factors that will not be altered automatically when water supply at the system level is augmented. As a result, the inequity in water allocation may actually increase if some households receive much more improvement in the services they receive after the completion of the water supply augmentation project than others.

Third, the augmentation in water supply through publicly funded infrastructure projects, such as the MWSP, may crowd out informal sector investment in urban water supply. In many cities in developing counties, water vendors play a critical role in water supply and distribution as piped water systems fail to provide adequate services to households (Whittington *et al.*, 1991; Kjellén & McGranahan, 2006). Without appropriate water tariffs in place to signal the real costs of the piped water and sanitation services, an increase in new water supplies may crowd out informal sector actors, such as water vendors, who often provide a critical backstop, reducing system vulnerabilities in water supply delivery.

Fourth, the water supply augmentation may accelerate a transition toward high household water consumption patterns that may not be sustainable in the long run. In the case of Kathmandu, for example, more households can be connected to the piped water system after the Melamchi project becomes fully operational, and households will use more water when the current water rationing is reduced. Fueled by low water tariffs, the water supply augmentation may unleash a significant increase in water use, which may aggravate the future water supply and demand balance and further complicate decisions about capacity expansion investments.

An appropriate policy framework, institutional arrangements, and management practices must be in place to avoid such pitfalls, but in many cities in developing countries, the weakness in water governance is a leading cause for water scarcity in the first place. While political opposition is often regarded as the main obstacle for changes in the policy framework, institutional arrangements, and management practices, the analytical challenge is equally daunting (Kandulu, 2013). Problems in urban water supply can be considered as 'wicked problems' which are subject to complex interdependence relationships among a diverse range of factors, and their resolution typically requires behavioral changes by both households and water utility managers (Lach *et al.*, 2005; Grafton, 2017).

Systems thinking can be critical in dealing with wicked problems. While water utility managers focus their work on piped water systems, their decisions and actions have significant impacts on the entire water system, including groundwater and the informal sector (Kjellén & McGranahan, 2006; Foster & van Steenbergen, 2011). The changes that will result in the whole system from the water augmentation project need to be analyzed; otherwise, achievement or progress made in one area may aggravate problems in other areas, a defining characteristics of wicked problems.

Behavioral science can also make significant contributions to improving water resource management in developing countries (Correia & Roseta-Palma, 2012). While most applications in this area so far have been focused on nudging and water-saving behavior (Merrett, 2002; Brick *et al.*, 2018), the behavioral insights may also make a significant contribution to the tariff designs and water allocations (Nauges & Whittington, 2019).

Summary of papers in the special issue

The six papers in this special issue reveal the challenges posed by rising water scarcity and the resulting housing-water tradeoffs in all of their complexity. In the first paper in this special issue, Chinnasamy and Shrestha note that in recent years KUKL, private households, and water vendors have been extracting more groundwater from the groundwater aquifer in the Kathmandu Valley than its sustainable yield. Their results show that the total annual groundwater extraction for the Kathmandu Valley is about 69 million cubic meters annually and that the drawdown of the deep aquifer has been on the order of 15–20 m since the early to mid-1980s.

Chinnasamy and Shrestha conclude that there has been substantial overexploitation of groundwater resources while Kathmandu has been waiting for the arrival of new water supplies from outside the Kathmandu Valley. The authors believe that when the MWSP is finished, it should provide sufficient new water supplies to reverse this decline in groundwater depletion. Their results suggest that the overexploitation of the groundwater resource in the Kathmandu Valley during the several decades spent waiting for out-of-basin supplies to arrive may, in fact, have been a reasonable strategy in the face of water scarcity. But for other cities facing water scarcity, this strategy may not be feasible, because out-of-basin supplies may not be available or affordable.

In the second paper, Thapa and his coauthors also address the groundwater situation in the Kathmandu Valley. They assess the spatial distribution of groundwater availability under five scenarios that represent conditions before and after the implementation of the MWSP. They use a numerical groundwater flow model with available local data and analyze water demand, supply infrastructure, changes in the hydraulic head (i.e. groundwater levels), groundwater pumping rates, and aquifer characteristics.

During the period 2003–2014, Thapa *et al.* find that the average hydraulic head in the wells in the valley declined from 2.57 meters below ground level (bgl) (0.23 meters per year) to 21.58 meters bgl (1.96 meters per year). Their model simulations suggest that changes in the average hydraulic head would range from +2.83 to +5.48 m at various stages of the MWSP implementation and −2.97 m for increased pumping rates if the MWSP was not implemented. The authors argue that the regulation of groundwater pumping is important for achieving the recovery of the valley's groundwater levels. They suggest that appropriate policy options could include groundwater extraction charges. They argue that such extraction charges should differ by location depending on the availability of new and existing public water supply in different areas. Based on their modeling work, the authors believe that the completion of the MWSP is an important positive step on the Kathmandu water development path.

In the third paper, Raina and coauthors paint a detailed picture of the complex water vending industry that has arisen in Kathmandu to provide households and businesses with water supplies to meet growing water demands. Their research shows that at the time of their study (2014), water vendors in the

Kathmandu Valley operated a diverse, heterogeneous group of businesses. There was a supply chain with two main products: bulk water and bottled drinking water. Transactions occurred in four main markets: two upstream markets (between water source owners and tanker trucks and between bottled water vendors and distributing vendors) and two consumer markets (between tanker truck vendors and consumers and between distributing vendors and consumers). Approximately 20 per cent of the water that households and businesses use in the Kathmandu Valley is delivered by this water vending supply chain.

However, Raina and coauthors show that these water vending services are not cheap. In the dry season, end users of vended water (both tanker truck water and bottled water) pay approximately 3.4 times as much for vended water as they pay for water from the public piped water distribution system (which is heavily subsidized). Nevertheless, the authors concluded that the portions of the water vending supply chain that they examined were all quite competitive.

The large amounts of money flowing through the water vending supply chain illustrate the revenue potential associated with improving services offered by the public piped distribution system. When the MWSP is completed, KUKL should be able to improve the quantity and quality of the water supplied through the piped water system. The authors' results suggest that most households in Kathmandu are already paying much more to water vendors than to KUKL, and that the completion of the MWSP offers KUKL a rare opportunity to raise water tariffs simultaneously with the delivery of improved piped services.

In the fourth paper, Chindarkar and her coauthors analyzed the relationship in Kathmandu between household time allocation, the reliability of tap water, and water consumption patterns under the current conditions of intermittent supply from the public piped system. Unsurprisingly, they find that time spent on productive activities is negatively correlated with time spent on water collection for the person most responsible for water collection in the household. Yet surprisingly, they also find that when tap water connections become more reliable, households spend more time for collecting water. As a consequence of this increased time investment, households consume more water both from their private tap and from all sources. Because an increase in reliability leads to an increase in collection time inside the home, the authors argue that when supplies are intermittent, it is important for the public water provider to arrange water delivery schedules to minimize disruptions that can affect work, leisure, and other activities. For cities with intermittent supply, the authors conclude that more research is needed to understand how individuals and households prioritize different activities in order to identify which times of day would be the most conducive to water collection.

In the fifth paper, Leong and Tan use qualitative research methods to explore the narrative structures that residents of Kathmandu use to describe their emotions about the delay in the completion of the MWSP and the consequences for their daily lives. Despite the long delay in the completion of the MWSP and the high cost this delay imposes on the public, economic and social life in Kathmandu continues to be vibrant. Based on 55 interviews conducted in Kathmandu in 2016, Leong and Tan explore people's subjective narratives about the delayed MWSP, in particular the self-reported perceptions of their resilience. Leong and Tan argue that people's ability to cope with adversity engenders the self-perception of resilience. They examine the paradox that the act of coping with water scarcity may *itself* create a false sense of resilience among the population.

Leong and Tan uncover three main themes in people's narratives. First, people in Kathmandu clearly understand that there is a high cost to the current situation. Participants were often acutely aware of the monetary cost of obtaining bottled and tanker water, and that the piped water supply has become more and more intermittent. Their respondents could easily cite figures and compare their coping costs with household income. Respondents also described changes to daily routines to obtain water and heated arguments at public taps. Yet despite the high costs, participants have adapted, accepting these costs as part of their daily lives.

The second theme to emerge from Leong and Tan's analysis is people's feeling that the current situation is unsustainable, and that the problem has (and will) worsen over time. The third theme relates to a perception of incompetence on the part of the government and generates the most emotional responses. Participants expressed anger at the incompetence and corruption within municipal government and KUKL, and bitter resignation at the lack of progress on the issue of water supply. Leong and Tan's research thus emphasizes the interrelationships between self-perceptions of resilience, the incremental worsening nature of the water scarcity problem, people's perceived ability to pay increasing coping costs, and a lack of confidence in government capacity.

In the sixth paper in this special issue, Suwal and coauthors examine the preferences of households in Kathmandu for different attributes of the tariff structure. They note that only rarely do water utility managers have information about what types of tariff structures households prefer. These authors find that many households express a preference for an increasing block tariff when it is described to them. However, when they are asked about what a fair water bill would be for a specified quantity of water, households support a tariff structure that determines household water bills as a linear function of water use, not a tariff structure that results in household water bills increasing as a nonlinear function of water use, such as an increasing block tariff. These results suggest that after the MWSP is finished and new water supplies arrive in Kathmandu, utility managers may have considerable latitude in choosing a tariff structure other than an IBT. A post-Melamchi tariff structure should be able to focus on objectives, such as cost recovery and economic efficiency, and allow affordability concerns to be addressed by customer assistance programs (Cook *et al.*, 2020a, 2020b).

Acknowledgement

We thank the Institute of Water Policy, Lee Kuan Yew School of Public Policy, and the National University of Singapore for their financial support in making this special issue possible.

References

Arnold, B. F., Null, C., Luby, S. P. & Colford, J. M. (2018). Implications of WASH benefits trials for water and sanitation – authors' reply. *The Lancet Global Health 6*(6), e616–e617.

Asian Development Bank (2013). *Kathmandu Valley Wastewater Management Project: Report and Recommendations of the President, Financial Analysis*. Manila, Philippines. Available at: https://www.adb.org/projects/documents/kathmandu-valley-wastewater-management-project-rrp (Accessed April 4 2017).

Asian Development Bank (2015). *Kathmandu Valley Water Supply Improvement Project – Additional Financing: Report and Recommendations of the President, Financial Analysis*. Manila, Philippines. Available at: https://www.adb.org/projects/documents/kathmandu-valley-water-supply-improvement-project-additional-financing-rrp (Accessed April 4 2017).

Bakker, K., Kooy, M., Shofiania, N. E. & Martijn, E.-J. (2008). Governance failure: rethinking the institutional dimensions of urban water supply to poor households. *World Development 36*(10), 1891–1915.

Brick, K., De Martino, S. & Visser, M. (2018). *Behavioural Nudges for Water Conservation: Experimental Evidence From Cape Town*. Draft Working Paper (February), pp. 1–58.

Brown, J., Albert, J. & Whittington, D. (2019). Community-led total sanitation moves the needle on ending open defecation in Zambia. *American Journal of Tropical Medicine and Hygiene 100*(4), 767–769.

Coffey, D. & Spears, D. (2018). Implications of WASH benefits trials for water and sanitation. *The Lancet Global Health 6*(6), e615.

Cook, J., Fuente, D. & Whittington, D. (2020a). Choosing among pro-poor policy options in water supply and sanitation. *Water Economics and Policy*. Forthcoming.

Cook, J., Whittington, D., Fuente, D. & Matichich, M. (2020b). A global assessment of non-tariff customer assistance programs in water supply and sanitation. In: *Development Studies in Regional Science*. Chen, Z., Bowen, W. M. & Whittington, D. (eds). Manila, Springer, Forthcoming.

Correia, R. & Roseta-Palma, C. (2012). Behavioural economics in water management. In: *12th Biennial Conference of the International Society for Ecological Economics: ISEE 2012: Ecological Economics and Rio +20: Contributions And Challenges For A Green Economy*, Rio de Janeiro.

Cumming, O. & Curtis, V. (2018). Implications of WASH benefits trials for water and sanitation. *The Lancet Global Health 6*(6), e613–e614.

Cutler, D. & Miller, G. (2005). The role of public health improvements in health advances: the twentieth-century United States. *Demography 42*(1), 1–22.

Esrey, S. A. (1996). Water, waste, and well-being: a multicountry study. *American Journal of Epidemiology 143*, 608–623.

Foster, S. & van Steenbergen, F. (2011). Conjunctive use of groundwater: a 'lost opportunity' for water management in the developing world? *Hydrogeology Journal 19*(5), 959–962.

Grafton, R. Q. (2017). Responding to the 'wicked problem' of water security. *Water Resources Management 31*, 10.

Grafton, R. Q., Chu, L. & Kompas, T. (2015). Optimal water tariffs and supply augmentation for cost-of-service regulated water utilities. *Utilities Policy 34*(C), 54–62.

Gurung, Y., Zhao, J., Kumar, B., Suwal, B., Wu, X. & Whittington, D. (2017). The costs of delay in infrastructure investments: a comparison of 2001 and 2014 household water supply coping costs in the Kathmandu Valley, Nepal. *Water Resources Research 53*, 7078–7102. doi:10.1002/2016WR019529.

Kandulu, J. (2013). Are we risking too much? Estimating the risk-mitigating benefit of urban water supply augmentation options. *Water Supply 13*(2), 513–540.

Kjellén, M. & McGranahan, G. (2006). *Informal Water Vendors and the Urban Poor*. IIED, London (HSDP Series, Theme: Water-3).

Lach, D., Rayner, S. & Ingram, H. (2005). Taming the waters: strategies to domesticate the wicked problems of water resource management. *International Journal of Water 3*(1), 1–17.

Merrett, S. (2002). Behavioural studies of the domestic demand for water services in Africa. *Water Policy 4*(1), 69–81.

Muller, A. M. (2017). *Understanding Cape Town's Water Crisis*. Civil Engineering. June 2017. Available at: https://ssrn.com/abstract=2995937

Nauges, C. & Whittington, D. (2019). Social norms information treatments in the municipal water supply sector: some new insights on benefits and costs. *Water Economics and Policy 5*(3), 1–40.

Pattanayak, S. K., Yang, J. -C., Whittington, D. & Kumar, K. C. B. (2005). Coping with unreliable public water supplies: averting expenditures by households in Kathmandu, Nepal. *Water Resources Research 41*, W02012.

Pickering, A. J., Null, C., Winch, P. J., Mangwadu, G., Arnold, B. F., Prendergast, A. J., Njenga, S. M., Rahman, M., Ntozini, R., Benjamin-Chung, J., Stewart, C. P., Huda, T. M. N., Moulton, L. H., Colford Jr., J. M., Luby, S. P. & Humphrey, J. H. (2019). The WASH benefits and SHINE trials: interpretation of WASH intervention effects on linear growth and diarrhoea. *The Lancet Global Health 7*(8), e1139–e1146.

Sahin, O., Siems, R. S., Stewart, R. A. & Porter, M. G. (2016). Paradigm shift to enhanced water supply planning through augmented grids, scarcity pricing and adaptive factory water: a system dynamics approach. *Environmental Modeling and Software 75*, 348–361.

Saleth, M. & Dinar, A. (1997). *Satisfying Urban Thirst Water Supply Augmentation and Pricing Policy in Hyderabad City, India*. World Bank Technical Paper No. 395.

Whittington, D., Lauria, D. T. & Mu, X. (1991). A study of water vending and willingness to pay for water in Onitsha, Nigeria. *World Development 19*(2/3), 179–198.

Whittington, D., Pattanayak, S. K., Yang, J. C. & Kumar, K. B. (2002). Household demand for improved piped water services: evidence from Kathmandu, Nepal. *Water Policy 4*(6), 531–556.

Whittington, D., Radin, M. & Jeuland, M. (2020). Evidence-based policy analysis? The strange case of the randomized controlled trials of community-led total sanitation. *Oxford Economic Policy Review*. Forthcoming.

Households' preferences for water tariff structures in Kathmandu, Nepal

Bhim Raj Suwal[a], Jane Zhao[b], Aditi Raina[c], Xun Wu[d], Namrata Chindarkar[d], K. C. Bal Kumar[a] and Dale Whittington[e,f,g,]*

[a]*Central Department of Population Studies, Tribhuvan University, Kathmandu, Nepal*
[b]*Department of City & Regional Planning, University of North Carolina, Chapel Hill, NC, USA*
[c]*Lee Kwan Yew School of Public Policy, National University of Singapore, Singapore*
[d]*Division of Public Policy, Hong Kong University of Science and Technology, Hong Kong*
[e]*Departments of Environmental Sciences & Engineering and City & Regional Planning, University of North Carolina, Chapel Hill, NC, USA*
[f]*Global Development Institute, University of Manchester, Manchester, UK*
[g]*Institute of Water Policy, Lee Kwan Yew School of Public Policy, National University of Singapore, Singapore*
**Corresponding author. E-mail: Dale_Whittington@unc.edu*

Abstract

Despite being politically sensitive, water tariffs are frequently administered without information about households' preferences for tariff structures. In this paper we examine the tariff preferences of 1,500 households in Kathmandu, Nepal. We first use a bivariate probit model to examine stated preferences for (1) an increasing block tariff (IBT) and (2) a positive fixed charge. We find that household preferences for IBTs and fixed charges are not easily explained by household socioeconomic and water use characteristics. Second, we ask respondents what they think a fair water bill would be for a randomly assigned quantity of water. We model the responses as a function of both quantity and household socioeconomic and water use characteristics. While households support a water tariff that results in a household's water bill increasing as a household's water use increases, we do not find evidence that households support an increasing, non-linear relationship between water use and a household's water bill. Our results suggest that respondents desire affordable piped water services and water bills that are calculated fairly for everyone. Because the notion of fairness in Kathmandu varies, utility managers may have considerable latitude in choosing a tariff structure that focuses on other objectives, such as cost recovery, revenue stability, and economic efficiency.

Keywords: Household preferences; Increasing block tariff; Tariff design; Water tariff structure

doi: 10.2166/wp.2019.079

Introduction

Water utility managers, their tariff consultants, and regulators rarely consult a utility's customers about the form of tariff structure households would prefer. Both citizens and water utility managers pay a great deal of attention to changes in fixed charges and volumetric prices charged to customers, but much less attention to customers' preferences for the tariff structure itself. Water tariff design is typically assumed to be best left to experts who can most appropriately balance the utility's objectives of cost recovery and revenue stability with their customers' objectives of equity, fairness, affordability, and economic efficiency.

Because in most locations water utilities operate as monopoly suppliers of piped water and sewer services, their tariff structures are effectively administered prices, and utility customers have few opportunities to reveal their preferences for alternative pricing structures[1]. As a result, there can be a serious mismatch between the water tariff structure used and household preferences for the tariff structure. Unpopular tariff structures can have a long life because customers have no easy way to drive them out of the market. This is quite different, for example, from the telecommunications sector where competition can be intense. Mobile phone companies are consistently offering customers a variety of different rate plans from which they can choose, and unpopular rate plans quickly disappear from the market.

There is a sizeable consulting business in the preparation of tariff studies for water utilities, and there is near global consensus among water utility managers and international donors that increasing block tariffs (IBTs) are the tariff structure best suited to balancing the multiple objectives of tariff design, and ensuring that poor households have access to sufficient quantities of water at affordable prices. Consultants specializing in advising water utilities on water tariff structures and price levels typically assume that an IBT should be used, and then discuss with utility management the pros and cons of such issues as the number and size of blocks, the size of any positive fixed charges, minimum monthly water bills, and seasonal pricing.

Numerous scholars, however, have challenged this global consensus on IBTs, arguing that IBTs neither target subsidies to poor households effectively nor send the correct signals about the economic value of water (Boland & Whittington, 2000; Foster & Araujo, 2004; Komives *et al.*, 2007; Diakité *et al.*, 2009; Angel-Urdinola & Wodon, 2012; Whittington *et al.*, 2015; Fuente *et al.*, 2016; Nauges & Whittington, 2017). Given this debate between scholars and practitioners over the efficiency and effectiveness of IBTs, it is timely to ask what households themselves think. In fact, utilities rarely elicit input from households on tariff design issues. In some countries, public hearings might be held to gauge citizens' reactions to specific proposals from tariff consultants, but such input is often sought after the main tariff design issues have already been decided.

There are several reasons that household preferences should be considered in the selection of a water tariff structure. First, households may have a preference for a specific water tariff structure, and the utility's choice of tariff structure may affect their well-being. From a cost–benefit perspective, ignoring household preferences will miss an opportunity to increase welfare. Second, households are probably more likely to pay water bills and to support a water utility that uses a tariff structure that they believe is fair and efficient. Perceptions of fairness are important for the cooperation of households with an institution (Pokharel, 2015). Third, many households likely believe that they have a right to have their voices heard in matters

[1] In Scotland, some households are now able to choose their bulk water supplier, but globally this is very rare.

of pricing such basic services as water. Ignoring household preferences on the tariff structure deployed may expose water utilities and policy-makers to greater political risks (Crase *et al.*, 2008).

Water tariff structures are often designed based on certain assumptions about household behaviors and how such behaviors may be altered by price mechanisms. Such assumptions should be based on a careful investigation of households' preferences. For example, IBTs are often proposed based on the assumption that they will provide the right 'price signal' to consumers. In fact, this marginal price signal can only be provided to households in one of the blocks of the IBT (most likely the highest price block).

In this paper we report the results from a survey conducted in 2014 of 1,500 households in Kathmandu, Nepal, in which we asked respondents for their opinions about different water tariff structures. Specifically, we asked households about their preferences regarding three attributes of a water tariff: (1) its structure (IBT vs uniform volumetric pricing); (2) positive fixed charges; and (3) household flexibility in selecting rate plans. We collected information on households' knowledge, attitudes, and perceptions about the water services in Kathmandu; and households' urban environmental and infrastructure priorities. We also asked respondents what they thought a 'fair water bill' would be for a randomly assigned fixed quantity of water. We analyzed these answers to determine whether respondents had a preference for increasing nonlinear tariff structures. We find that respondents support monthly water bills that increase linearly as the quantity of water use increases.

In the next section of the paper, we summarize the prevalence of different tariff structures being used globally and then present the tariff structure currently used in Kathmandu. The third section describes our fieldwork in Kathmandu and provides a socioeconomic profile of our sample respondents. The fourth section describes our analytical strategy. In the fifth section we present the results. We first describe households' knowledge, attitudes, and perceptions about the water supply in Kathmandu, and the priority they place on reducing water shortages compared to other urban problems. We then present our results regarding households' preferences for different attributes of water tariff structures, as well as regression analyses that examine the types of households that prefer different tariff structures. The sixth section summarizes our results and offers concluding remarks.

Background

Most water utilities globally now use IBTs to calculate customers' water bills. The Global Water Intelligence (GWI) database reported 122 out of 165 selected utilities from around the world were using IBTs in 2013. China had been one of the last refuges of uniform volumetric tariffs, but in 2013, the Chinese government mandated that all cities in China must adopt IBTs by the end of 2015 (National Development and Reform Commission, 2013). The predominance of IBTs globally can be expected to grow as Chinese cities make the shift to IBTs over the next few years.

Although there is a clear preference for the basic IBT structure, there is no consensus on the details. IBTs have four main features: (1) number of blocks; (2) size of the blocks, especially the first ('lifeline') block; (3) presence and size of a positive fixed charge; and (4) the volumetric prices charged in each block. Tariff consultants and utility managers argue over these features, and there is wide variation across all four globally.

For example, the number of blocks in the IBTs in the GWI database varies widely. Of the 122 IBTs, the most common number of blocks was three (28%), followed by four blocks (21%). However, 6% had two blocks, 11% had five blocks, 10% had six blocks, 7% had seven blocks, and 18% had eight blocks.

The GWI database also reports sizes of the first (lifeline) block for the IBTs. The median size is 10 m^3 in East Asia, South Asia, and Latin America, and somewhat lower in North and sub-Saharan Africa (8 and 7 m^3 per month, respectively), but the standard deviation is large. With such large lifeline blocks, it is not usual for some utilities to have a large proportion, even a majority, of their residential consumers falling into the lifeline block.

Positive fixed charges raise the average price paid by small volume-using households more than that of large volume-using households and thus penalize the same households that the IBT's lifeline block is designed to help. Nevertheless, most water utilities in the GWI database that use IBTs (70%) add a positive fixed charge to the volumetric component of their customers' water bills. The mean positive fixed charge was US$4.35 per month. Only 30% of utilities with uniform volumetric tariffs add a fixed charge, and their average fixed charge is lower (US$3.20 per month)[2], which suggests that more of these uniform tariff-setting utilities may be aware of the adverse effect of fixed charges on poor households.

One would expect volumetric prices in different blocks to vary across utilities due to differences in the costs of water and wastewater services, and the differences are indeed substantial. However, across all of the IBT structures, volumetric prices are very low in all of the blocks (not just the lifeline block). Considering only the 34 IBTs with three blocks (the most common IBT in the GWI database), the median volumetric price was US$0.35 per cubic meter in the first block, US$0.57 per cubic meter in the second block, and US$0.75 per cubic meter in the third block.

Most water utilities subsidize water use, and the average cost of water supply is more than the price in the highest-price block. Thus, the more water a household uses, the higher the subsidy it receives (Nauges & Whittington, 2017).

Kathmandu case study

The Kathmandu water utility (Kathmandu Upatyaka Khanepani Limited, KUKL) uses an IBT with a minimum bill. KUKL currently has about 190,000 residential customers (KUKL, 2015). The water bills of households with both metered and unmetered connections to the pipe network are calculated based on the pipe size of their connection. Most connected households have a half-inch pipe connection. Households with a metered one-half inch pipe connection are charged 100 NPR (US$0.96) for the first 10 cubic meters. For additional water use above the minimum 10 cubic meters, a household is charged 32 NPR (US$0.31) per cubic meter. For example, a household with a metered half-inch pipe connection that used 18 cubic meters a month would receive a water bill of 356 NPR (100 NPR for the first 10 cubic meters plus 32 NPR per cubic meter for the subsequent 8 cubic meters). Households with half-inch, unmetered connections are charged 785 NPR (US$7.50) per month.

This tariff structure has the appearance of an IBT because, if the household uses all 10 cubic meters, its average price is just 10 NPR (≈US$0.10) per cubic meter, and for additional water use the price is 32 NPR per cubic meter. However, the minimum bill of 100 NPR means that the marginal price for additional water use within the 'lifeline' block of 10 cubic meters is effectively zero because households

[2] The median positive fixed charges in East Asia (US$5) and Latin America (US$4 per month) are much higher than those in sub-Saharan Africa, the Middle East, North Africa, and South Asia (all about US$1 per month).

must pay this minimum charge regardless of how much water they use (e.g., they are charged the same 100 NPR whether they use 5, 7, or 10 cubic meters).

The implementation of an infrastructure investment program (the Melamchi Water Supply Project) is now underway. When completed, it will provide a new raw water supply for the Kathmandu Valley, and improvements in both the water and wastewater networks, as well as wastewater treatment facilities. Donor support for this project had been conditional upon tariff reform, and originally consultants had estimated that a 13-fold increase of the water tariff would be required between 1999 and 2009 to meet operating costs of the water network (Domènech *et al.*, 2013). The revised 2009 tariff increased the tariff set in 2004 by 10–30%, depending on the amount consumed (Himalayan Times, 2009; ADB, 2013). It included a lifeline block, but NGOs deemed this insufficient to assist poor households and demanded targeted subsidies for the poor (Gutierrez *et al.*, 2003; Domènech *et al.*, 2013). A new tariff structure was introduced in 2013, which increased the average tariff by 82%. This price increase, however, was still not enough to cover operating and maintenance costs, service charges, and debt (ADB, 2015). Further changes in the current tariff structure are thus a possibility, and our questions about tariff preferences were highly salient to sample respondents.

Fieldwork and socioeconomic profile of sample respondents

Our survey of 1,500 households was conducted in five municipalities in the Kathmandu Valley (Kathmandu, Lalitpur, Bhaktapur, Kirtipur, and Madhyapur) from August to October 2014. Households in this 2014 survey were selected based on a 2001 sample that was drawn to be representative of the population of Kathmandu at the time (Gurung *et al.*, 2017). We chose to re-interview the households from 2001 in order to construct a panel dataset. The sample is thus not representative of the 2014 population of households (differences are reported in the Appendix, Table A1, available online). In 2001, households were selected using a multi-stage clustered sampling procedure (see Whittington *et al.* (2002) and Pattanayak *et al.* (2005) for more details). In the 2014 re-survey of the 2001 households, if we were unable to locate the original household, a nearby household in the same cluster was selected for an interview[3]. In total, we were able to locate and re-interview 927 of the 1,500 households in the 2001 survey. In the 2014 survey, there are thus 573 replacement households[4].

Table 1 presents a socioeconomic profile of the sample households. As shown, monthly mean household income was 70,441 NPR (about US$700). The average household size was five members. The average age of the head of household was 57 years (reflecting the sampling strategy of re-surveying the 2001 sample). Table 1 also reports summary statistics on infrastructure services and housing. Virtually every household in the sample had electricity (99.7%) and some sort of access to the telecommunications network (98%), either via a landline (64%) or a cellphone. Households had, on average,

[3] If the household head from the 2001 sample household was missing, the next most senior member of the house was interviewed.

[4] Although the 2001 households were a representative sample of the population in the five municipalities in Kathmandu Valley, the sample from the 2014 re-survey is not because many households migrated to Kathmandu over the period from 2001 to 2014. These households are not part of our 2014 sample unless they happened to be included as a replacement household. Appendix Table A1 shows that the survey households are more likely to be homeowners, to use LP gas as the cooking fuel, to have a flush toilet connected to the sewer systems, to own more durable goods, and are more likely to be literate in Kathmandu and Lalitpur.

Table 1. Socioeconomic profile of sample households.

Variable	N	Median	Mean	SD
Monthly household income	1,500	57,500	70,441	74,299
Household size	1,500	5	5.05	2.12
Head of household age	1,500	56	56.52	12.86
Electricity	1,500	1	1.00	0.05
Landline	1,500	1	0.64	0.48
Number of cellphones	1,495	3	3.20	1.64
Electricity bill	1,476	700	897	749
Phone bill	1,500	1,000	1,288	1,178

three cellphones. Mean monthly phone bills were 1,288 NPR (US$13), about 2% of monthly income. Average monthly electricity bills were 897 NPR (US$9), about 1% of monthly income.

Table 2 presents some quantitative characteristics of the piped water system, as reported by our respondents. Seventy per cent of the sample households had a private water connection. Households with private connections reported that they received very little water from their connection during the dry season (median and mean are 75 and 126 L/household/day, respectively), and about twice that during the rainy season (median and mean of 143 and 240 L/household/day, respectively). Households received water from their private connections for only about an hour to an hour and a half, on average, once every 5 days. Most households (59%) reported having a working water meter, and 67% received a water bill (a large majority reporting receiving their bill monthly). The median monthly bill of a household was 150 NPR (US$1.55), with a mean water bill of 218 NPR (US$2.25), 0.5% of monthly household income. The average price paid for the water from a private connection was 170 NPR per m^3 (US$1.75 per m^3); the median price was 71 NPR per m^3 (US$0.73 per m^3).

Analytical strategy for investigating households' preferences for water tariff structures

We examine households' preferences for different tariff structures in two ways. First, we asked respondents a series of direct questions. Enumerators explained to each of the households in the

Table 2. Piped water summary statistics.

	N	Median	Mean	SD
Private water connection (1 = yes, 0 = no)	1,500	1	0.70	0.46
Water use – rainy season (L/day)	1,046	143	240	329
Water use – dry season (L/day)	988	75	126	183
Days between service	1,051	5	4.99	2.93
Length of service each time water comes (min)	1,050	60	98.6	84.2
Hours of service per month	1,050	7.5	11.5	14.2
Water meter (1 = yes, 0 = no)	1,500	1	0.63	0.48
Working water meter (1 = yes, 0 = no)	1,500	1	0.59	0.49
Received water bill (1 = yes, 0 = no)	1,500	1	0.67	0.47
Monthly water bill	1,051	150	218	229
Average price paid for water from the piped connection (NPR/m^3)	988	71	170	337

2014 survey how water bills are calculated when a water utility uses either a uniform volumetric tariff or an IBT. Respondents were not given specific volumetric prices for the two tariff structures, or other details about the IBT structure, such as the number and size of blocks.

Respondents were then asked which tariff structure they would prefer the water utility to adopt and the reasons why they selected their preferred tariff structure. Enumerators next asked respondents whether their choice of tariff structure was based on whether it was best for their household or all the households in Kathmandu.

Enumerators also explained that many water utilities used a two-part tariff that added a positive fixed charge to the volumetric component to calculate customers' water bills. Respondents were asked if they thought the use of a positive fixed charge was a good idea. Next, enumerators reminded respondents that mobile phone customers could choose the rate plan (tariff structure) that best suited their household. Enumerators asked respondents if they would like to be able to choose their tariff structure for piped water and wastewater services (even if the respondent's household did not have a piped connection).

We checked respondents' understanding of the IBT concept. Those who did not pass the check question (e.g., 'IBT means that the price per unit of water will increase in stages as a household's total water use increases. True or False?') are not included in all parts of the analysis ($N = 271$ from 1,500 households). Understanding of the IBT concept is most critical for the choice between an IBT and uniform volumetric tariff structure; it is not as important for respondents' choice of a fixed charge[5].

The propensity to select an IBT and the propensity to select a fixed charge are both dichotomous variables. Therefore, a latent variable model is used to examine their relationships with each other and socioeconomic factors, water use, and other preferences. Because these decision variables could be connected, unobservable variables may affect both the propensity to choose an IBT and a fixed charge. The seemingly unrelated probit model jointly estimates these two propensities with their correlated disturbances (Greene, 2011). We test for exogeneity using maximum-likelihood simultaneous estimation of the two probit equations, i.e., recursive bivariate probit (Maddala, 1983; Costa-Font & Gil, 2005). This model allows us to correct for some of the unobserved heterogeneity that can contribute to omitted variable bias and is also expected to increase the efficiency of the estimation (Costa-Font & Gil, 2005; Greene, 2011).

The specification for our two-equation model is as follows:

$$IBT_i^* = \beta_0 + \beta_1 SE_{1i} + \varepsilon_{1i},\ IBT_i = 1 \text{ if } IBT_i^* > 0,\ 0 \text{ otherwise,} \tag{1}$$

$$Fixed_i^* = \delta_0 + \delta_1 SE_{2i} + \varepsilon_{2i},\ \ Fixed_i = 1 \text{ if } Fixed_i^* > 0,\ 0 \text{ otherwise,} \begin{pmatrix} \varepsilon_{1i} \\ \varepsilon_{2i} \end{pmatrix} | SE_{1i},\ SE_{2i} \sim N\left[\begin{pmatrix} 0 \\ 0 \end{pmatrix}, \begin{pmatrix} 1 & \rho \\ \rho & 1 \end{pmatrix}\right] \tag{2}$$

where IBT_i = binary indicator for household i's stated preference for a uniform volumetric or IBT, $Fixed_i$ = binary indicator for household i's stated preference for a fixed charge, SE_{1i}, SE_{2i} = vector of socioeconomic characteristics of household i, ε_{1i}, ε_{2i} = random error terms.

[5] Appendix Table A2 shows the differences between these two sub-samples. Respondents who responded 'not sure' or 'do not know' are also excluded (N = 22). A selection model was considered but the likelihood ratio test of independent equations failed to be rejected, suggesting there is no selection bias due to a respondent initially failing to pass the check question.

IBT_i^*, $Fixed_i^*$ are latent variables observed as dummy variables IBT_i, $Fixed_i$, respectively. SE_{1i}, SE_{2i} are exogenous variables, and β_0, β_1, δ_0, δ_1 are parameters of the preference functions. The error terms of the two equations are modeled to be dependent and distributed as a bivariate normal. The Wald test for $\rho = 0$ provides evidence about the correlation between the error terms of the equations and if the models should therefore be jointly estimated.

Second, in a different section of the questionnaire, households were also asked the following question:

After the Melamchi water supply project is completed, households with piped water connections will have better service. What do you think a fair[6] *monthly water bill would be for a household that uses [5000, 10000, 15,000, 20,000, 25,000, 30,000] liters per month?*

The different quantities of monthly water use (5,000, 10,000, 15,000, 20,000, 25,000, 30,000 liters per month; or 5, 10, 15, 20, 25, 30 m^3 per month) were randomly assigned to sample respondents. Each respondent answered this question only once for the single randomly assigned quantity that they received. We estimated the relationship between the respondent's reported 'fair monthly water bill,' the exogenously assigned quantity and socioeconomic covariates of the respondent's household socioeconomic characteristics using the following equation:

$$\text{WaterBill}_i = \beta_0 + \beta_1 Q_i + \beta_2 (Q_{ij})^2 + \beta_3 SE_i + \varepsilon \tag{3}$$

where WaterBill_i = self-reported fair water bill of household i, $Q_{i,j}$ = quantity of water j randomly assigned to household i (j = 5,000, 10,000, 15,000, 20,000, 25,000, 30,000 liters per month), SE_i = vector of socioeconomic characteristics of household i.

We expect that respondents, in aggregate, believe it would be fair for a household that uses more water to pay a higher water bill, so $\beta_1 > 0$. We are agnostic about the sign and significance of β_2. If respondents, in aggregate, believe that a household that uses more water should pay an increasingly higher water bill, then $\beta_2 > 0$. For example, if households favor an IBT or other form of increasing non-linear tariff structure, then we expect $\beta_2 > 0$. Similarly, if households favor a decreasing block or other decreasing nonlinear tariff structure, $\beta_2 < 0$. If households favor a uniform volumetric tariff, we expect β_1 to be positive, but for β_2 not to be statistically significant.

This model describes aggregated household preferences. Households were not asked to state fair monthly water bills for multiple hypothetical quantities, so we cannot use these data to test whether the respondent understood how the household's water bill would be calculated if an IBT were used. Even if all households preferred and understood an IBT, they may still disagree on sizes and prices of blocks, and thus β_1 and β_s might not necessarily be significant and/or positive.

For all three models, fixed effects at the municipality level are included. Standard errors are clustered at the neighborhood level.

---

[6] We note that there is no direct translation for the word 'fair' in Nepali. There are two closely related words in Nepali that reflect two closely related concepts – fairness in process and fairness in outcome (Pokharel, 2015). However, these concepts do not apply precisely to the task of expressing a single estimate for a 'fair' water bill associated with a specified quantity of water. However, we believe that respondents' answers to the question posed reflect an understanding of the English meaning of 'fair water bill'.

Results

Salience of the water supply situation in Kathmandu

We first demonstrate that the water shortage in Kathmandu was highly salient to sample respondents. Table 3 reports respondents' answers to a question that asked their first, second, and third priorities from a predetermined list of environmental and infrastructure problems. Respondents were most concerned about three urban infrastructure problems: (1) water shortage, (2) electricity outages, and (3) poor garbage collection and solid waste management. Of these three, addressing the water shortage was overwhelmingly respondents' top priority.

This conclusion is reinforced by respondents' answers to another question in the survey. We told respondents in households with and without private connections to:

> '... *Suppose that you are able to have potable water from a private connection and the water service is available several hours per day, seven days per week, and that your monthly water bill will be equal to or less than what you are currently paying for water from all sources. In which of the following areas would you like to adjust your water use: drinking, cooking, bathing, cleaning house, outdoor gardening?*'

The closed-end answers were, 'increased a little,' 'increased substantially' and 'no change.' Table 4 shows that the majority of respondents expect that their household's water use would increase a little or substantially for three large components of household water uses: bathing, cleaning house, and outdoor gardening.

Preferences for IBT vs uniform volumetric, fixed charges, tariff flexibility at the household level

We next examine reported preferences for tariff structures, fixed charges, and tariff flexibility at the household level. We also examine water infrastructure and socioeconomic characteristics of households

Table 3. Urban environmental and infrastructure priorities*.

	First	Second	Third
Water shortage	57%	21%	9%
Electricity outage	14%	39%	23%
Air pollution	9%	5%	10%
Poor garbage collection and solid waste	8%	15%	27%
Contamination of drinking water	6%	7%	9%
Poor sewerage	5%	8%	9%
Improper disposal of hazardous wastes	1%	1%	2%
Water pollution in rivers	1%	2%	5%
Poor drainage/flooding	0%	2%	2%
Too much noise	0%	1%	2%

*Survey question: Which one of these environmental problems is the (first/second/third) most important that the government should solve in this city?

Table 4. Anticipated changes in water use after completion of Melamchi Project (number of respondents).

	Increase a little	Increase substantially	No change
1. Drinking	234	31	1,235
2. Cooking	409	55	1,036
3. Bathing	808	309	383
4. Cleaning house	820	318	362
5. Outdoor gardening	765	290	445
6. Washing clothes	53	32	1,415

that are associated with their tariff preferences. We first report the descriptive statistics and then the results of our probit models.

Table 5 compares the characteristics of households who prefer the IBT to a uniform volumetric rate. Fifty-eight percent of households (705 households) expressed a preference for the IBT compared to the uniform volumetric tariff (505 households). Twenty-five per cent of those who preferred the IBT also preferred a positive fixed charge. Most of those who prefer a uniform volumetric tariff (59%) also prefer a positive fixed charge. Additionally, 70% of all households preferred to have some choice of their water tariff structure.

Most respondents (78%) said that their choice of tariff structure was based on what they thought would be best for everyone in Kathmandu, not just themselves; 22% said that they were thinking

Table 5. Household tariff preferences.

	Preferred IBT	Preferred uniform volumetric tariff	Total
Tariff choice (number of households)	705	505	1,210
Preferred positive fixed charge	25%	59%	39%
Tariff flexibility	70%	70%	70%
Considered other households in Kathmandu in tariff choice	84%	70%	78%
Current water use from tap (m^3/month)	1.67	2.82	2.15
Reasons for tariff choice	My household gets a discount if we use less water (42%)	Same price rate for all customers (42%)	
	People who consume more water should pay higher prices for the extra water they consume (30%)	Encourages water conservation (21%)	
	Encourages water conservation (18%)	Cost of supplying water is the same so price should be the same (20%)	
	Helps poor households (10%)	Easy to calculate expenses on piped water (13%)	
What do you like least about the service from the piped water connection?	Less than 24-hour service (39%)	Less than 24-hour service (59%)	
	Unfair distribution (31%)	Poor water quality (19%)	
	Poor water quality (21%)	Unfair distribution (16%)	

about their own household. Those who preferred the IBT were more likely to report that they based their decision on what was good for everyone in Kathmandu (84%) compared to those who preferred the uniform tariff (70% said they based their decision on what was good for everyone in Kathmandu).

We then asked the reasons for preferring one tariff structure over the other. Those who chose the IBT said that they like the fact that they could 'get [a] discount if [they] use less water' (42%), suggesting that the price of water in the higher-priced blocks may have been perceived as the default price. They also believe that 'people who consume more water should pay [a] higher price for the extra water they consume' (30%). Those who chose the uniform tariff said that they liked the fact that it is the 'same price for all customers' (42%), that it '[encourages] water conservation' (21%), and that because 'the cost of supplying water is the same, the price should also be the same' (20%).

For those who answered that they were thinking mostly about their own family, we asked if their tariff structure preference would change if they thought about all households instead of just about themselves. Twenty-three per cent of those who initially chose the IBT stated they would change their answer. Similarly, 21% of those who initially chose the uniform tariff also said that they would change their answer. In contrast, only about 2% of those who said they based their decision on what was good for everyone in Kathmandu stated they would change their answer if they were thinking only about their own family.

Finally, we asked respondents if those with a connection to the piped water network should be able to choose the tariff plan that they want, just as cell phone users are. A majority of respondents (68%) responded yes, households should be able to choose, while 18% said households should not be allowed to, and 14% of respondents were unsure.

Table 6 presents the summary statistics for the dependent variables used in Equations (1)–(3), the key socioeconomic characteristics of the household (income, household size, respondent's education, head of the household responding), variables related to water usage (monthly expenditures, quantity of water collected from the tap), water shortage being the most important environmental problem, and other independent variables of interest. The literature on determinants of household preferences for IBTs is limited and provides little guidance on appropriate model specifications and covariates. Crase *et al.* (2007) find that households that supported an IBT are more likely to report a larger number of activities that use water. They also include socioeconomic variables as co-variates but find no significant association between household support for an IBT and age, household size, household income, or the size of the last water bill.

Table 7 shows the relationship between the stated preferences for an IBT and a fixed charge. Forty-two per cent of the sample ($N = 1{,}167$) prefer an IBT but not a fixed charge; 25% prefer a fixed charge and a uniform tariff instead of an IBT. Seventeen per cent do not want a fixed charge and do not prefer an IBT; 15% prefer both a fixed charge and an IBT. The tetrachoric correlation coefficient is estimated at −0.509, with an asymptotic standard error of 0.038, showing that those who prefer an IBT are less likely to prefer a fixed charge.

Table 8 reports the results of the bivariate probit model of preferences for IBTs and fixed charges. The first specification for both models looks only at socioeconomic controls: income, household size, respondent education, and if the respondent is the head of the household. Models (1) and (2) are estimated simultaneously. Model (1) shows that for household preferences for IBTs, income and household size do not have significant coefficients. More educated respondents are more likely to choose an IBT, and respondents that are household heads are less likely to choose an IBT. There are also municipality effects; households living in Kirtipur and Madyapur Thimi are more likely to choose an IBT compared to households living in Kathmandu (municipality). Model (2) shows that for household preferences for fixed charges, household income, and head of household have statistically significant

Table 6. Summary statistics.

Variable	Variable description	N	Mean	SD	Min	Max
ibt	Increasing block is best for KUKL to adopt (1 = yes, 0 = no)	1,210	0.58	0.49	0	1
fixed	Favors fixed charge in water bill (1 = yes, 0 = no)	1,439	0.40	0.49	0	1
fair_bill	Fair water bill post-Melamchi (NPR/month)	1,283	528	637	100	10,000
hyp_usage	Exogenously assigned hypothetical household usage (L)	1,500	16,150	8,118	5,000	30,000
hyp_usage_sq	Hypothetical household usage squared	1,500	3.3×10^8	2.8×10^8	2.5×10^7	9.0×10^8
ln_income	ln(Total household reported income)	1,450	10.92	0.76	6.21	13.92
nhh	Household size	1,500	5.05	2.12	1	15
resp_edu	Highest level of education of respondent (years), with illiteracy coded as zero	1,500	8.38	6.22	0	18
resp_illiterate	Respondent is illiterate (1 = yes, 0 = no)	1,500	0.219	0.413	0	1
resp_hhhead	Respondent is household head (1 = yes, 0 = no)	1,500	0.529	0.499	0	1
ln_monthlywaterexp	ln(Perceived monthly expenditures on water (NPR))	1,500	725	921	0	9,130
pwc_m3	Estimated quantity of water collected from piped water connection (m^3/month, dry season)	1,500	2.49	4.81	0	90
priority_water	Water shortage is first most important environmental problem (1 = yes, 0 = no)	1,500	0.57	0.50	0	1
consider_ktm	When choosing preferred tariff, thinking about what is best for everyone in Kathmandu (1 = yes, 0 = no)	1,500	0.76	0.42	0	1
change_mind	Answer would change if thinking about other group (either own hh or everyone) (1 = yes, 0 = no)	1,500	0.07	0.25	0	1

Table 7. Stated preferences for IBTs and fixed charges.

	No fixed charge	Fixed charge	Total
Uniform	201	297	498
IBT	494	175	669
Total	695	472	1,167

coefficients at the 1% and 10% levels, respectively. The higher the household income, the less likely the household is to prefer a fixed charge. Heads of households are more likely to prefer a fixed charge. Household preferences for fixed charges do not have a statistically significant relationship with household size, respondent education, or municipality.

The second set of specifications in Table 8 adds covariates related to water usage: monthly water expenditures (*ln_monthlywaterexp*), quantity of water collected from the private tap (*pwc_m3*), and if water shortage is the most important environmental problem (*priority_water*). Models (3) and (4) are estimated simultaneously. In model (3), the coefficient on quantity of water collected from the private tap (*pwc_m3*) is significant at the 1% level and is negative. The more water the respondent's household obtains from the private water connection, the less likely they were to choose an IBT. The addition of these water usage variables does not change the sign or significance of other variables, except for size of household (*nhh*), which becomes significant at the 10% level. In model (4), the coefficients on the water usage variables are not statistically significant.

Table 8. Biprobit regressions of preferences for IBTs and fixed charges.

Variables	(1) ibt	(2) fixed	(3) ibt	(4) fixed	(5) ibt	(6) fixed
ln_income	0.0710 (0.0582)	−0.158*** (0.0583)	0.0800 (0.0589)	−0.156** (0.0620)	0.0859 (0.0575)	−0.156*** (0.0604)
nhh	−0.0327 (0.0213)	0.0139 (0.0176)	−0.0365* (0.0210)	0.0155 (0.0174)	−0.0395* (0.0224)	0.0178 (0.0174)
resp_edu	0.0227*** (0.00723)	0.00384 (0.00966)	0.0237*** (0.00693)	0.00371 (0.0100)	0.0221*** (0.00719)	0.00407 (0.0100)
resp_illiterate	0.252** (0.119)	−0.00805 (0.127)	0.263** (0.115)	−0.0148 (0.126)	0.234* (0.121)	−0.00697 (0.126)
resp_hhhead	−0.175** (0.0744)	0.159* (0.0861)	−0.189** (0.0776)	0.166** (0.0845)	−0.204** (0.0797)	0.176** (0.0860)
ln_monthlywaterexp			0.0394 (0.0340)	−0.0329 (0.0228)	0.0480 (0.0331)	−0.0360 (0.0224)
pwc_m3			−0.0445*** (0.0136)	0.00811 (0.0108)	−0.0498*** (0.0134)	0.00984 (0.0113)
priority_water			−0.0693 (0.102)	−0.0901 (0.110)	−0.0378 (0.0983)	−0.105 (0.109)
consider_ktm					0.590*** (0.142)	−0.243 (0.175)
change_mind					0.203 (0.160)	0.136 (0.207)
Municipality – Lalitpur	0.270* (0.158)	−0.0353 (0.150)	0.254 (0.165)	−0.0490 (0.146)	0.205 (0.160)	−0.0173 (0.155)
Municipality – Bhaktapur	0.210 (0.160)	0.183 (0.136)	0.318 (0.225)	0.105 (0.141)	0.277 (0.243)	0.101 (0.137)
Municipality – Kirtipur	0.768*** (0.137)	−0.139 (0.165)	0.895*** (0.195)	−0.251 (0.167)	0.886*** (0.199)	−0.222 (0.171)
Municipality – Madhyapur Thimi	0.629*** (0.205)	0.0934 (0.162)	0.613*** (0.211)	0.111 (0.176)	0.588*** (0.217)	0.147 (0.180)
Constant	−0.724 (0.640)	1.292** (0.600)	−0.895 (0.665)	1.480** (0.648)	−1.423** (0.631)	1.654*** (0.641)
Observations	1,135		1,135		1,135	
Rho	− 0.518***		− 0.514***		− 0.506***	
Wald test of rho = 0 (*p*-value)	0.000		0.000		0.000	
AIC	2,916		2,905		2,873	
Wald Chi-square	170.1		225.7		403.8	
Correct classification (%)	43.6%		45.2%		45.3%	

Robust standard errors in parentheses.
$^{***}p < 0.01$, $^{**}p < 0.05$, $^{*}p < 0.1$.

The third set of specifications includes two additional variables: (1) whether the respondent was thinking about her own household or all of Kathmandu (*consider_ktm*) and (2) if they would change their mind (*change_mind*) when considering the other group instead (either own household or all of Kathmandu). Models (5) and (6) are estimated simultaneously. In model (5), the coefficient on whether the respondent considers all of Kathmandu when selecting the preferred tariff structure (*consider_ktm*) is positive and significant at the 1% level. When the respondent considers the larger community, they are more likely to select an IBT over a uniform volumetric tariff. In model (6), we find that neither considering all of Kathmandu nor changing their mind have statistically significant effects on household preferences for a fixed charge.

Across all specifications, the value of the correlation between the error terms of the IBT and fixed charge equations (*rho*) and its significance level are reported. All specifications show *rho* values that are significant at the 1% level, rejecting the null hypothesis that $rho = 0$ and that the error terms are uncorrelated. Comparing the specifications, we look at the Akaike information criteria (AIC). The last specification reports the lowest AIC of 2,873 and is therefore the preferred specification. The Wald Chi-square test statistics are sufficiently high for all models to reject the null hypothesis that all coefficients in the model are zero. However, the ability for the three specifications to correctly classify both outcomes is low, ranging from 43.6% to 45.3%.

The models describing household preferences for IBTs and fixed charges are relatively weak, with low predictive abilities. However, we do find that households that are more likely to be large users (larger household and higher monthly water consumption) are less likely to favor IBTs. This is not surprising if households are rational economic actors and are cost-minimizing. Additionally, considering what is good for Kathmandu as a whole has a large and significant coefficient, suggesting that households' intuition and opinions agree with common pro-IBT arguments (Boland & Whittington, 2000). Paired with the qualitative questions about reasons for preferring one tariff structure over another, these findings make sense. It would seem that ideas about fairness or social justice (i.e., 'good for everyone in Kathmandu'), cost-minimization, and water conservation drive household preferences, and are not easily approximated using household socioeconomic and water use characteristics.

Preferences for nonlinear tariff structures: fair water bills

Figure 1 presents the frequency distributions of respondents' reported fair water bills at each of the exogenously assigned monthly quantities of water use. These frequency distributions show that reported fair water bills gradually shift higher as the quantity of water increases, but many respondents reported very low 'fair water bills' at all quantities. There was also a great deal of heterogeneity in reported fair water bills at each exogenously assigned monthly quantity of water use.

Table 9 describes respondents' reported fair water bills and implicit average water prices at different hypothetical monthly quantities of water, with references to what the water bill and average water price would be under the current KUKL water tariff structure. For the overall sample, the fair water bill is monotonically increasing as a function of hypothetical water use. The mean fair water bill is higher than what the household would pay under the current tariff for all hypothetical water use levels. The correlation coefficient between fair water bill and the hypothetical monthly water quantity is positive but low (Pearson's is 0.27 and Spearman's is 0.38). Many respondents report a very low 'fair water bill' at all quantities. The implicit average water price is monotonically decreasing in volume of water use (except for the mean

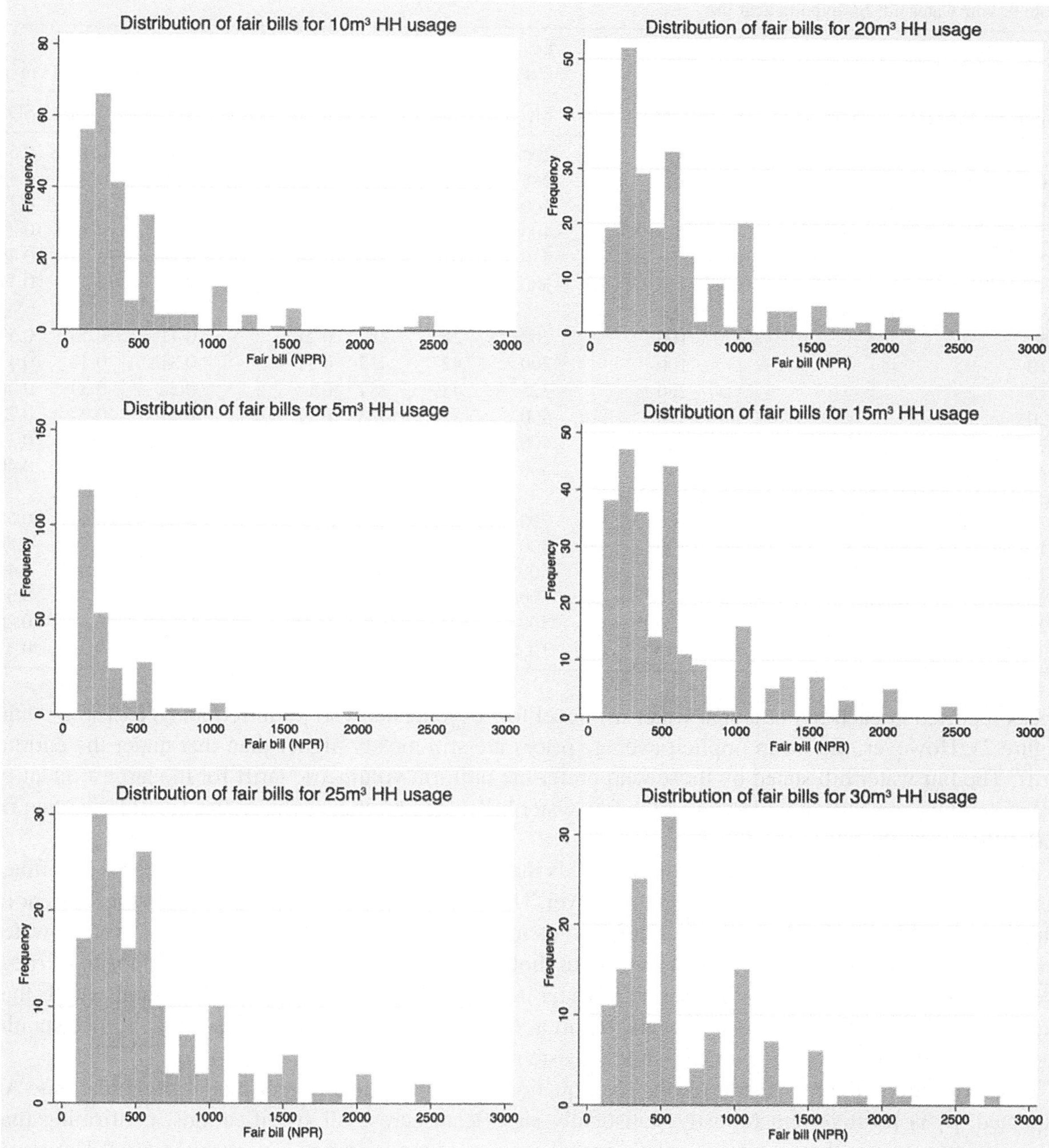

Fig. 1. Distribution of responses for fair monthly water bill by exogenously assigned hypothetical household usage.

price at 30 m^3/month). Again, the mean implicit average water price is also much higher than what the household would pay under the current tariff – more than double for 5, 10, and 15 m^3.

Comparing the frequency distributions of reported fair water bills for those who preferred an IBT with those who prefer a uniform volumetric tariff, we see lower fair water bills and implicit average water

Table 9. Fair water bill by hypothetical use.

			Fair water bill (NPR/month)			Implicit average water price (USD/m^3)			
Hypothetical water use (m^3/month)	N	Current tariff	Median	Mean	SD	Current tariff	Median	Mean	SD
5	246	100	200	272	270	0.21	0.41	0.56	0.56
10	244	100	288	423	438	0.11	0.30	0.44	0.45
15	247	260	400	546	542	0.18	0.27	0.38	0.37
20	224	420	400	580	518	0.22	0.21	0.30	0.27
25	168	580	450	680	979	0.24	0.19	0.28	0.40
30	154	740	500	835	909	0.26	0.17	0.29	0.31
Prefer increasing block tariff									
5	120	100	200	269	287	0.21	0.41	0.55	0.59
10	115	100	300	427	432	0.11	0.31	0.44	0.45
15	114	260	325	489	382	0.18	0.22	0.34	0.26
20	105	420	400	520	433	0.22	0.21	0.27	0.22
25	81	580	400	610	779	0.24	0.16	0.25	0.32
30	65	740	500	686	862	0.26	0.17	0.24	0.30
Prefer uniform volumetric tariff									
5	88	100	200	303	282	0.21	0.41	0.62	0.58
10	84	100	300	486	513	0.11	0.31	0.50	0.53
15	86	260	500	673	600	0.18	0.34	0.46	0.41
20	76	420	500	752	669	0.22	0.26	0.39	0.34
25	51	580	500	728	629	0.24	0.21	0.30	0.26
30	49	740	1000	1144	891	0.26	0.34	0.39	0.31

prices reported at each hypothetical water use level for respondents who preferred an IBT (Table 9 and Figure 2). However, the mean implicit average prices are still mostly higher than that under the current tariff. The fair water bill stated by those who prefer the uniform volumetric tariff for the largest quantity (30 m^3/month) is nearly double that of the fair water bill stated by those who prefer an IBT (median of 500 NPR vs 1,000 NPR).

The implicit average water price for households that stated they preferred an IBT decreases as a function of hypothetical water use at the aggregate level. This reflects an inconsistency between the household and aggregated levels. While individual households may state that they support increasing average water prices as a function of water use, when these households' responses are aggregated, the opposite is true – average water prices decrease as a function of water use. The results in Table 9 show that households that express a preference for IBTs, in aggregate, report a 'fair water bill' that implies the implicit price should decrease as volume increases (which is not consistent with an IBT structure).

Table 10 reports the regression results for Equation (3), with four different model specifications. As expected, β_1 is positive and robustly statistically significant across all specifications, confirming that respondents perceive that it is fair for households who use more water to pay more in total. However, β_2 is negative but not statistically significant, suggesting that respondents do not prefer increasing non-linear tariff structures such as IBTs. The last model specification includes preferences for IBTs and the interaction between *ibt* and *hyp_usage_sq*. The coefficient on *ibt* is not statistically significant, but the coefficient on the interaction term is. This implies that a preference for IBTs modifies β_2 negatively, which is the opposite of what we would expect. For those who believe households that use more water should pay increasingly higher water bills, β_2 should be positive. Household income is positively

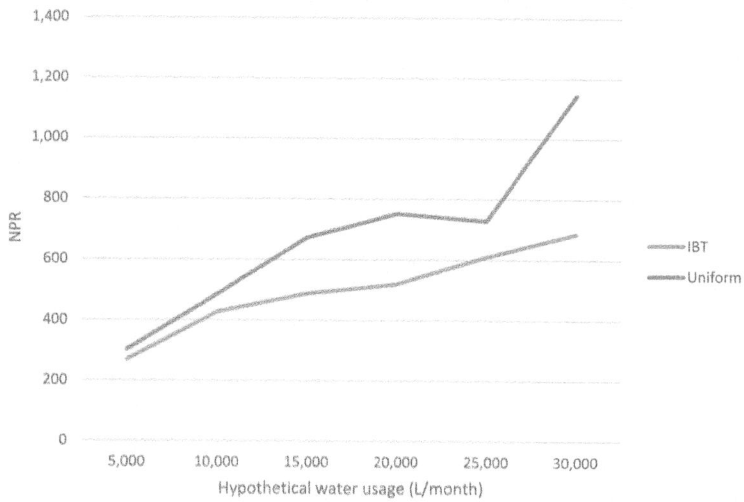

Fig. 2. Fair water bill (NPR).

associated with higher reported fair water bills, as are higher monthly water expenditures. The R^2 values for the model specifications range from 0.12 to 0.18[7].

Our analysis suggests that respondents do not have strong preferences for IBTs or other increasing nonlinear tariff structures. Even though a small majority of respondents (58%) said that they prefer IBTs to uniform volumetric prices in response to a direct question, after controlling for household characteristics and neighborhood unobservables, we find little evidence that households prefer increasing nonlinear tariffs. We believe that many respondents simply wanted cheap, affordable piped water services, and that water bills should be calculated fairly for everyone in Kathmandu, which they do not necessarily associate with an IBT or other increasing nonlinear tariff structure.

Water shortage was clearly a salient problem, with strong, explicitly stated dissatisfaction regarding the level of service and the fairness of distribution. Increases in household water use would occur in bathing, cleaning, and gardening. Household preferences also reflect a strong desire for more affordable water. Households are currently receiving little water from their taps and paying an average of US\$1.75/m^3 (SD 3.47). Implicit average water prices from stated fair bills are significantly lower (see Table 9).

We found that preferences for tariff structure are difficult to associate with socioeconomic variables. Instead, it seems that preferences are driven by desires for cost-minimization and fairness. When it comes to IBTs, the variable with the strongest explanatory power is consideration of what is best for other households. It also seems that households who use less water understand that they will pay less under an IBT. Overall fairness is a large concern, but the interpretation of fairness varies. Those who prefer an IBT believe

[7] Appendix Table A3 presents results for the same set of specifications, but separately for those who prefer an IBT and for those who prefer a uniform volumetric tariff. We find that the relationship between quantity and price (β_1) holds for those who prefer a uniform volumetric tariff, but find no significant relationship (β_1 and β_2) for those who prefer an IBT. Our results are not sensitive to the exclusion of respondents who did not pass the check question on understanding of the IBT concept. Appendix Table A4 presents the results for the entire sample. The sign and significance of key coefficients β_1 and β_2 remain robust. The magnitude of β_1 remains stable at 0.02 across all specifications. Appendix Table A5 modifies the model by including a dummy variable for the initial failure of the check question (*check_understanding*). While the coefficient on the response to the check question is statistically significant, the sign and significance of β_1 and β_2 remain the same.

Table 10. Regressions of fair bill on hypothetical monthly usage.

Variables	(1) fair_bill	(2) fair_bill	(3) fair_bill	(4) fair_bill
hyp_usage	0.0256** (0.0105)	0.0225** (0.0106)	0.0238** (0.0109)	0.0242** (0.0108)
hyp_usage_sq	-1.44×10^{-7} (3.33×10^{-7})	-6.45×10^{-8} (3.37×10^{-7})	-1.04×10^{-7} (3.50×10^{-7})	7.93×10^{-8} (3.19×10^{-7})
ln_income		121.0*** (38.71)	101.0*** (36.27)	106.5*** (35.57)
nhh		−6.776 (9.062)	−9.864 (9.389)	−12.51 (9.036)
resp_edu		1.547 (4.296)	1.088 (4.235)	1.935 (4.468)
resp_illiterate		31.25 (52.18)	35.81 (53.39)	47.61 (53.62)
resp_hhhead		97.02*** (31.74)	86.27*** (30.93)	73.97** (30.21)
ln_monthlywaterexp			32.21*** (10.71)	35.00*** (9.053)
pwc_m3			6.425 (4.389)	3.045 (3.761)
priority_water			−57.20 (35.86)	−57.43 (34.45)
ibt				−55.45 (55.15)
ibtXhyp_usage_sq				-3.21×10^{-7}* (1.68×10^{-7})
Municipality – Lalitpur	−112.3 (67.76)	−86.07 (65.44)	−66.31 (64.42)	−49.13 (58.90)
Municipality – Bhaktapur	−288.1*** (62.37)	−215.1*** (67.57)	−121.2 (98.56)	−96.27 (93.22)
Municipality – Kirtipur	−346.9*** (58.04)	−296.8*** (50.92)	−213.1*** (58.88)	−161.8*** (46.51)
Municipality – Madhyapur Thimi	−154.1** (75.91)	−91.77 (76.43)	−88.40 (76.41)	−53.98 (69.96)
Constant	251.9*** (73.83)	−1,102*** (372.8)	−1,041*** (362.4)	−1,079*** (349.4)
Observations	1,040	1,011	1,011	1,005
R^2	0.116	0.140	0.157	0.177
F-statistic	18.17	12.81	12.00	13.28
Prob > F	0.000	0.000	0.000	0.000

Clustered standard errors in parentheses.
***$p < 0.01$, **$p < 0.05$, *$p < 0.1$.

that people who use less can get a discount and those who consume more should pay higher prices for the extra water they consume. Those who prefer a uniform tariff believe that all customers should pay the same price.

From Figure 2 and Tables 9 and 10, there is evidence for a positive relationship between price and quantity of water. Fair water bills generally increase monotonically as a function of water use. However, subsample analysis (see Appendix Table A3) shows that the significant relationship between quantity and price is driven by those who prefer a uniform volumetric tariff. For those who prefer an IBT, neither β_1 nor β_2 are significant. While there is an overall positive relationship between water use and water price, there is little evidence in support of IBTs.

Conclusions

We find that households are primarily concerned with the perceived fairness of the water tariff structure. Based on their responses about 'fair water bills' for exogenously assigned quantities, households appear to support a water tariff that is a function of the quantity of water used. Importantly, the notion of fairness varies across the population. For those who prefer an IBT, fairness means that people who use less should get a greater discount per unit of water used than those who consume more. In contrast with this position, fairness among those who prefer a uniform tariff means that all customers should pay the same price regardless of the quantity of water they use.

Perhaps surprisingly, there is little evidence that households prefer that water tariffs be structured in a way that ensures poor households have access to sufficient quantities of water at affordable prices. While this is an important objective of consultants, water utility managers, and international donors, our sample households seem more concerned about the price level generally instead of the affordability of piped water services for poor households.

Our multivariate analysis illustrates how household preference for tariff structure can be analyzed and results interpreted. The simple descriptive statistics show that a small majority of respondents (58%) said that they prefer IBTs to uniform volumetric prices in response to a direct question, might lead to a 'naive' conclusion that IBTs are preferred by the most households. But after controlling for household characteristics and neighborhood unobservables, we find little evidence that households have strong preferences for increasing nonlinear tariffs.

Our findings have important implications for the choices over different water tariff structures. One reason IBTs are popular with water utility managers may be that they believe citizens consider IBTs to be fair. Water utility managers may be unaware that their customers may have alternative notions of fairness, and that many may not, in fact, favor IBTs. In most cities in low- and middle-income countries, the volumetric prices in all the blocks of an IBT are below the average costs of service. Our findings are consistent with the hypothesis that this may be due in part to the fact that there may be little customer support for high volumetric prices in the upper blocks.

Water utility managers should take household preferences into account when choosing a tariff structure. However, because household preferences for water tariff structures in Kathmandu are heterogeneous, utility managers may have considerable latitude in choosing a tariff structure that best achieves the utility's objectives of cost recovery and revenue stability. As long as the utility managers focus on and communicate the affordability and fairness of the water tariff, they should be able to garner public support for tariff reforms.

Finally, we suggest that there is a need for further empirical research on households' notions of fairness and social justice and how households assess the consequences of changes in both tariff structures and price levels in terms of these criteria. Additionally, as this study did not examine household preferences about sizes and prices of blocks in an IBT, future work is needed to more clearly understand both individual and aggregate preferences about more detailed features of IBTs.

Acknowledgments

We would like to express our thanks to Yogendra B. Gurung for his assistance with the design and management of the fieldwork, and to the Institute of Water Policy, National University of Singapore, for their financial support. We would also like to thank Michael Young, Céline Nauges, Meenu Tewari, Marc Jeuland, and Atul Pokharel for their constructive comments.

Supplementary material

The Supplementary Material for this paper is available online at http://dx.doi.org/10.2166/wp.2019.079.

References

Angel-Urdinola, D. F. & Wodon, Q. (2012). Does increasing access to infrastructure services improve the targeting performance of water subsidies? *Journal of International Development 24*(1), 88–101.

Asian Development Bank (2013). *Kathmandu Valley Wastewater Management Project: Report and Recommendations of the President, Financial Analysis*. Manila, Philippines. Available at: https://www.adb.org/projects/documents/kathmandu-valley-wastewater-management-project-rrp (accessed April 4 2017).

Asian Development Bank (2015). *Kathmandu Valley Water Supply Improvement Project – Additional Financing: Report and Recommendations of the President, Financial Analysis*. Manila, Philippines. Available at: https://www.adb.org/projects/documents/kathmandu-valley-water-supply-improvement-project-additional-financing-rrp (accessed April 4 2017).

Boland, J. & Whittington, D. (2000). The political economy of water tariff design in developing countries: IBTs versus uniform price with rebate. In: *The Political Economy of Water Pricing Reforms*. Dinar, A. (ed.). Oxford University Press, New York, USA.

Costa-Font, J. & Gil, J. (2005). Obesity and the incidence of chronic diseases in Spain: a seemingly unrelated probit approach. *Economics & Human Biology 3*(2), 188–214.

Crase, L., O'Keefe, S. & Burston, J. (2007). Inclining block tariffs for urban water. *Agenda: A Journal of Policy Analysis and Reform 14*(1), 69–80.

Crase, L., O'Keefe, S. & Dollery, B. (2008). Urban water pricing: practical perspectives and customer preferences. *Economic Papers: A Journal of Applied Economics and Policy 27*(2), 194–206.

Diakité, D., Semenov, A. & Thomas, A. (2009). A proposal for social pricing of water supply in Côte d'Ivoire. *Journal of Development Economics 88*(2), 258–268.

Domènech, L., March, H. & Sauri, D. (2013). Contesting large-scale water supply projects at both ends of the pipe in Kathmandu and Melamchi Valleys, Nepal. *Geoforum 47*, 22–31.

Foster, V. & Araujo, C. (2004). *Does Infrastructure Reform Work for the Poor? A Case Study From Guatemala*. World Bank Policy Research Working Paper No. 3185. World Bank, Washington, DC, USA.

Fuente, D., Gakii Gatua, J., Ikiara, M., Kabubo-Mariara, J., Mwaura, M. & Whittington, D. (2016). Water and sanitation service delivery, pricing, and the poor: an empirical analysis of subsidy incidence in Nairobi, Kenya. *Water Resources Research 52*(6), 4845–4862.

Greene, W. (2011). *Econometric Analysis. Seventh Edition*. Pearson Education Limited, Essex, UK.

Gurung, Y., Zhao, J., Kumar, B., Suwal, B., Wu, X. & Whittington, D. (2017). The costs of delay in infrastructure investments: a comparison of 2001 and 2014 household water supply coping costs in the Kathmandu Valley, Nepal. *Water Resources Research 53*, 7078–7102. doi:10.1002/2016WR019529.

Gutierrez, E., Calaguas, B., Green, J. & Roaf, V. (2003). *New Rules, New Roles: Does PSP Benefit the Poor?* WaterAid and Tearfund, London, UK.

Himalayan Times (2009). *Supplier Hikes Water Tariff*. 16 December 2009. Available at: https://thehimalayantimes.com/kathmandu/supplier-hikes-water-tariff/ (accessed 4 April 2017).

Kathmandu Upatyaka Khanepani Limited (2015). *Annual Report*. Kathmandu, Nepal.

Komives, K., Halpern, J., Foster, V., Wodon, Q. & Abdullah, R. (2007). Utility subsidies as social transfers: an empirical evaluation of targeting performance. *Development Policy Review 25*(6), 659–679.

Maddala, G. (1983). *Limited-Dependent and Qualitative Variables in Econometrics. Econometric Society Monographs No. 3*. Cambridge University Press, New York, USA.

National Development and Reform Commission (2013). *On Accelerating the Establishment of the Ladder Price System for Water for Urban Residents*. No. 2676. Ministry of Housing and Urban-Rural Development. Available at: http://www.ndrc.gov.cn/fzgggz/jggl/zcfg/201401/t20140103_574403.html.

Nauges, C. & Whittington, D. (2017). Evaluating the performance of alternative municipal tariff designs: quantifying the tradeoffs between equity, economic efficiency, and cost recovery. *World Development 91*, 125–143.

Pattanayak, S. K., Yang, J.-C., Whittington, D. & Kumar, K. C. B. (2005). Coping with unreliable public water supplies: averting expenditures by households in Kathmandu, Nepal. *Water Resources Research 41*, W02012.

Pokharel, A. (2015). *Towards a Theory of Sustained Cooperation*. Watson Institute for International Studies Research Paper No. 2015-25.

Whittington, D., Pattanayak, S. K., Yang, J. C. & Kumar, K. B. (2002). Household demand for improved piped water services: evidence from Kathmandu, Nepal. *Water Policy 4*(6), 531–556.

Whittington, D., Nauges, C., Fuente, D. & Wu, X. (2015). A diagnostic tool for estimating the incidence of subsidies delivered by water utilities in low- and medium-income countries, with illustrative simulations. *Utilities Policy 34*, 70–81.

Melamchi water supply project: potential to replenish Kathmandu's groundwater status for dry season access

Pennan Chinnasamy IWA[a,b,*] and Surendra Raj Shrestha[c]

[a]Indian Institute of Technology, Bombay, Mumbai, Maharashtra, India

*Corresponding author. E-mail: p.chinnasamy@iitb.ac.in

[b]International Water Management Institute, Sanepa, Lalitpur, Nepal

[c]Ground Water Resources Development Board (GWRDB), Ministry of Irrigation, Government of Nepal, Babarmahal, Kathmandu, Nepal

Abstract

Kathmandu Upatyaka Khanepani Limited (KUKL) currently uses 35 surface and 57 groundwater sources to supply water for Nepal's capital, Kathmandu. It is necessary to understand if the Melamchi Water Supply Project (MWSP) can assist lean period water supply by indirectly increasing groundwater storage, through diverting excess water supply to groundwater recharge zones. The current study analyzed long-term groundwater depletion to assess available groundwater storage, followed by assessment of groundwater balance for the Kathmandu Valley. Results show that total groundwater extraction for Kathmandu was 69.44 million cubic meters (MCM) and drawdown of the groundwater surface was 15–20 m since the construction of wells in 1984/85, indicating substantial overexploitation. Results indicate that the ongoing unmet demand of 170 MCM/year can be easily satisfied if groundwater storage is recharged effectively, as underground water storage potential is 246 MCM/year due to a groundwater depletion rate of 2–10 m. From results, it is evident that that the timely implementation of the MWSP can help ease ongoing water stress and aid in reversing the damage caused to groundwater storage. In the long run, MWSP can supply water and recharge groundwater during monsoon periods, thus improving the quality of life and socio-economic status in Kathmandu.

Keywords: Groundwater; Kathmandu; Melamchi Water Supply Project; Nepal; Water supply

Introduction

Nepal is one of the fastest urbanizing countries of South Asia, with Kathmandu, the capital city, having the highest urbanization rate in the nation (Jha & Shrestha, 2010). As a result, the city faces water management issues in supplying for the water demands of the growing city (Lacombe *et al.*, 2019). Kathmandu houses the largest population (approximately 1.5 million) and economy of Nepal

doi: 10.2166/wp.2019.080

with 50% of urban population and 30% of the national gross domestic product (HMG & IUCN, 1995; HMG & MoPE, 1999; ADB, 2010). The population of Kathmandu is growing at 4.5% per annum, and with migration and economic impacts, the growth rate is projected to increase further in the coming decades (MoPE *et al.*, 2001; Central Bureau of Statistics, 2011). With the population increase, there is considerable increase in stress on the Government of Nepal to provide domestic water supply (HMG & IUCN, 1995; HMG & MoPE, 1999; Metcalf & Eddy, 1999, 2000, 2002; HMG, 2004; Udmale *et al.*, 2016). In particular, drinking water quality and quantity is one of the biggest concerns in the Kathmandu Valley (Cresswell *et al.*, 2001; Chapagain, 2014). Even though Kathmandu is rapidly urbanizing, long-term water supply schemes to sustain such growth are limited, even before any population explosion (Binnie & Partners, 1973a, 1973b, 1989). The drinking water supply sector for the valley has been subjected to many reforms since 1891 (Chapagain, 2014). One key institutional reform was in the separation from the Nepal Water Supply Corporation (NWSC) and formation of the Kathmandu drinking water supply sector through the Kathmandu Valley Water Supply Management Board (KVWSMB). This also led to many policy reforms, in particular the introduction of Public Private Partnership (PPP), through which the Kathmandu Upatyaka Khanepani Limited (KUKL) was formed (ADB, 2000; Kathmandu Valley Water Supply Management Board, 2006; Chapagain, 2014).

The KUKL uses 35 surface water and 57 groundwater sources to supply water for Kathmandu (Jha *et al.*, 1997; GWRDB, 2012). According to KUKL the water demand in 2013 in the Kathmandu Valley is 360 million liters per day (MLD) but only 95–154 MLD of water is available, while 50% of it is obtained from the groundwater resources (KUKL, 2012). Stanley *et al.* (1994) estimated that sustainable groundwater withdrawal from the valley's aquifers is 26.3 MLD, while JICA (JICA, 1990a, 1990b, 1990c), suggested that groundwater can be drawn only at around 15 MLD. A well-inventory study by the Ground Water Resource Development Board in 2012 indicated the presence of 759 deep tube wells in Kathmandu. Total groundwater extraction by private and community tube wells is 31.15 million cubic meters (MCM) (GWRDB, 2012). The annual groundwater extraction by NWSC and KUKL wells in Kathmandu is 38.29 MCM (KVWSM, 2010). This shows the annual extraction of groundwater from the deep tube wells in Kathmandu is 69.44 MCM. Both analyses indicate a higher groundwater withdrawl rate than sustainable levels.

As a result of insufficient supply of surface water for both drinking and non-drinking uses, approximately 50% of the water supply in Kathmandu is from groundwater systems that comprise both shallow and deep aquifers (Jha *et al.*, 1997; Khatiwada *et al.*, 2002). Tube wells, dug wells, and stone spouts are the major mechanisms of groundwater use (ICON *et al.*, 2002). As a result, the groundwater of Kathmandu Valley is under immense pressure from heavy use for domestic and industrial purposes (Cresswell *et al.*, 2001). Under such conditions, and with limited water supply, Cresswell *et al.* (2001) indicated more than a decade ago that with the ongoing groundwater extraction rate, which is 20 times that of the recharge rate (15 mm per year), groundwater reserves will be empty in 100 years. Without sufficient institutional and policy reforms, the groundwater extraction rates are projected to increase across the valley. When this is combined with the population increase, the groundwater reserves will be emptied far earlier than the 100 year timeline.

Due to the unsustainable population growth and associated increase in water use, the citizens of Kathmandu Valley experience severe water stress for domestic use (Chapagain, 2014). Less than 50% of the water need is satisfied, thereby increasing stress on other water resources, especially groundwater to sustain livelihood and to invest for industrial purposes. However, as the groundwater resources are finite, the ongoing water stresses have led to unsustainable groundwater use. Under these circumstances, it is necessary to quantify the current groundwater trend in the region with an aim to understand if groundwater recharge activities can increase groundwater resources (e.g. Agoramoorthy *et al.*, 2016; Chinnasamy, 2016a; Chinnasamy & Agoramoorthy, 2016; Chinnasamy *et al.*, 2018a, 2018b; Jadeja *et al.*, 2018). In addition, there is a need to understand if the Melamchi Water Supply Project (MWSP) scheme will indirectly benefit groundwater resources if some of the excess water, especially during wet seasons, can be tactically used to recharge existing water harvesting structures (e.g. temple ponds, city ponds, wells, headwaters of ancient spouts, etc.).

Objectives

The major objective of this study is to assess the groundwater depletion trends for the Kathmandu Valley, using Ground Water Resources Development Board (GWRDB) records, with the aim of quantifying the increase in groundwater storage due to excess water supply from MWSP in groundwater recharge zones. The other objectives include assessment, from literature review, the potential of groundwater recharge zones and aquifer characteristics. With this information the authors aim to understand the enhanced groundwater recharge rates in the Kathmandu Valley due to the MWSP water supply, under the assumption that there may be an excess of water during wet seasons and leakage in water supply systems.

This paper, for the first time, will use observed groundwater data to monitor the current groundwater depletion in the Kathmandu Valley. The paper will also analyze the aquifer properties for the Kathmandu Valley aiming to estimate recharge rates. With the MWSP, it is estimated that the groundwater contribution for Kathmandu's water supply will be lower. However, due to the delay in the project and ever increasing population growth, groundwater resources have seen a decline beyond sustainable rates.

A combination of observed groundwater data, water supply/use data and groundwater recharge potential from MWSP (by using MWSP excess water to recharge existing ponds and lakes in the valley) will be analyzed to investigate sustainable methods for use of groundwater and surface water for Kathmandu. The results generated will illustrate the use of MWSP water during wet seasons and groundwater during dry seasons, thereby providing physical evidence to support the conjunctive use of groundwater and MWSP surface water for the Kathmandu Valley.

Methods

Study site

Kathmandu Valley, an oval-shaped tectonic basin, is one of the largest intermontane basins developed within the Lesser Himalaya of Central Nepal (Figure 1). The valley is 30 km long East to West and 25 km North to South with an area around 650 km^2. The Kathmandu Valley includes three districts, namely Kathmandu, Lalitpur and Bhaktapur, and is located between latitude 27,032′34″ and 27,049′11″ North, and longitude 85,011′10″ East and 85,031′10″ West. The valley is surrounded by the Mahabharat mountain range with four hills acting as forts of the valley: Phulchowki in the South East, Chandragiri/Champa Devi in the South West, Shivapuri in the North West, and Nagarkot in the North East. The Valley encloses the entire area of Bhaktapur district, 85% of Kathmandu district and 50% of Lalitpur district. The valley floor is typically plane with small hillocks, many river terraces and floodplains ranging in elevations between 1,300 and 1,400 m. The highest altitudes are 2,166 m

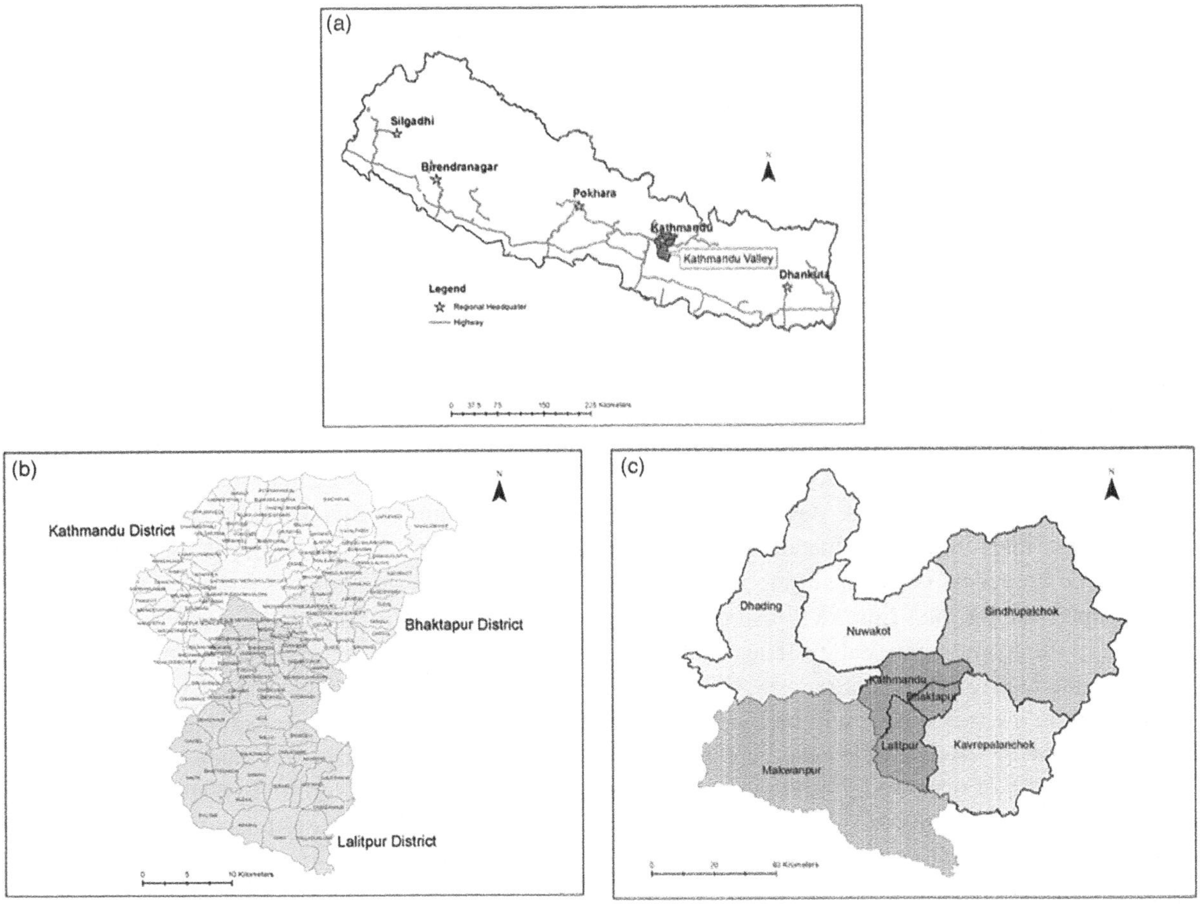

Fig. 1. Maps showing (a) location of Kathmandu Valley in Nepal (blue), (b) major districts in Kathmandu and (c) location of municipalities, and Village Development Committees (VDCs) in Kathmandu Valley. Please refer to the online version of this paper to see this figure in color: http://dx.doi.org/10.2166/wp.2019.080.

(in Bhaktapur), 2,732 m (in Kathmandu), and 2,831 m (in Lalitpur). Some of the higher mountain peaks surrounding the Kathmandu Valley are Phulchowki (2,800 m), Chandragiri (2,365 m), Nagarjun (2,100 m), Shivapuri (2,453 m) and Nagarkot (2,100 m) (GWRDB, 2012).

The Kathmandu climate is characterized as subtropical and cool temperate with four seasons: spring from March to May; summer from June to August; autumn from September to November; and winter from December to February. The rainy season is present during the monsoon period from June through September with about 80% of the annual rainfall (the average total precipitation in the Kathmandu Valley is 1,159.65 MCM per year) occurring during the rainy season. The rainfall varies with altitude with the higher elevations having higher rainfall than lower regions, for example only 1,300 mm per year rainfall occurs in the valley floor while 3,000 mm per year rainfall occurs in the mountain rim surrounding the valley. Average temperature for the region in summer season is around 30 °C and minimum temperature in winter season is about 0 °C (GWRDB, 2012).

The core of the Kathmandu Valley is densely populated with both urban and rural residents. The valley's population was estimated to be 3.5 million people in 2015, with the Kathmandu district having the largest share, followed by Lalitpur and Bhaktapur districts. Due to the continuing migration of rural population into the Kathmandu Valley, the expansion of rural areas into adjacent urban areas is likely to continue without regulation (GWRDB, 2012). KVWSWSI (2010) and Shrestha & Shukla (2010) predict that the population of the Kathmandu Valley will be 4.5 million and 5.8 million in 2020 and 2025, respectively.

Kathmandu Valley water supply and water stress

Kathmandu Valley's population is ever increasing due to migration of villagers for better employment. The KUKL is responsible for water supply in the region. The KUKL has been supplying water to the Kathmandu Valley from 35 surface water sources around the Kathmandu Valley peripheries, 57 deep tube wells (DTWs) and 11 dug wells. KUKL has 43 reservoirs and 20 treatment plants with a total treatment capacity of 117 MLD. The total water demand in year 2012 was 350 MLD. The production capacity of the KUKL system is 144 MLD in the wet season and 84 MLD in the dry season. The trend of water demand and supply scenarios in Kathmandu Valley is shown in Figure 2 and Table 1 (KUKL, 2012).

In addition to the KUKL water supply, most of the people depend on ancient water supply sources and groundwater resources (GWRDB, 2012). The valley has many traditional and ancient stone water spouts: around 237 in Kathmandu, 77 in Lalitpur, and 53 in Bhaktapur (Becker-Ritterspach, 1995; Amatya, 2006). These stone spouts were installed within rectilinear pits built into the ground and supplied through 'Raj Kulos' (state water canals) in a decentralized approach. In recent years, urbanization and the associated decline in water level across the valley have led to the drying up of the majority of these spouts.

Groundwater resources for water supply in the Kathmandu Valley

There has been immense stress on the groundwater resources of the Kathmandu region (Metcalf & Eddy, 1999, 2002; Shrestha *et al.*, 2013). The GWRDB recently conducted a study to assess the

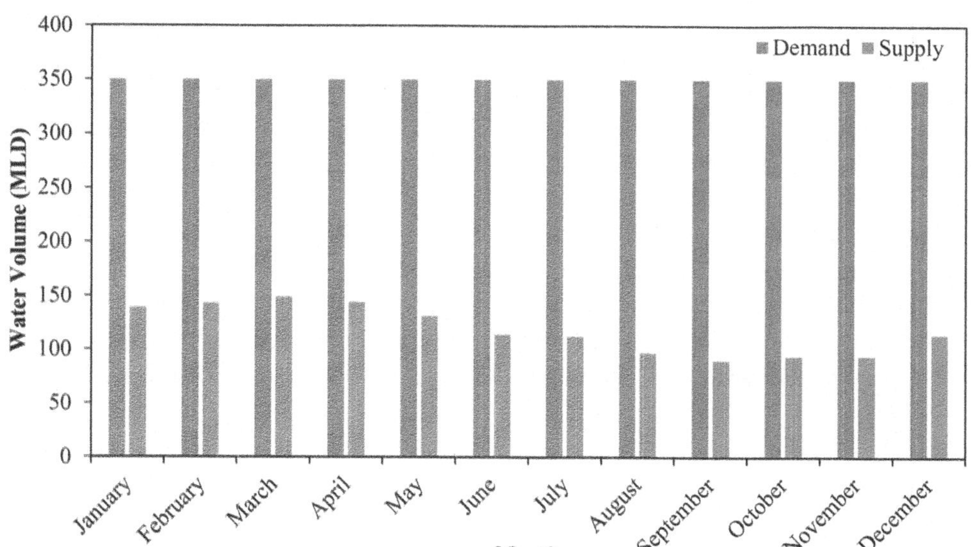

Fig. 2. Domestic water demand and KUKL water supply in the Kathmandu Valley (*Source*: KUKL, 2012).

Table 1. Water supply and demand in Kathmandu Valley (*Source*: KUKL, 2012).

Volume (MLD) /year	2010	2011	2012	2013
Water demand	320	350	350	360
Water supply	180	180	182	182
Unmet water demand	139	170	168	178

groundwater balance components and the water demand and supply scenario in the Kathmandu Valley (GWRDB, 2012). As per the report, approximately 20% of the water applied to irrigation is returned to the aquifers in the region. One important result from the study was that there is considerable leakage in the city's water supply, which indirectly recharges the local aquifers. The report estimates that despite the annual demand of 350 MLD water, NWSC/KUKL could only supply 182 MLD (KUKL, 2012). Of this total water supply, it is assumed that 54% of the water is supplied by using water from surface water sources, while the remaining 46% was supplied using water from deep tube wells (typical of depth from 50 to 200 meters). The report also indicated that 35–40% of the supplied water (49 MLD) is being lost due to leakage in the distribution system (KUKL, 2012). It is assumed that part of this water goes as baseflow, while the rest goes to recharge the groundwater aquifer of the valley.

Groundwater data

The groundwater monitoring by GWRDB was started from the year 2000 in the Kathmandu Valley, specifically covering the Melamchi water supply region. The monitoring wells were selected on the basis of available well inventory survey carried out by Melamchi Water Supply Development Board (MWSDB). In the beginning 50 wells were selected for the groundwater-level monitoring program. Out of the 50 selected GW level monitoring wells (both production and abandoned), the monthly monitoring works were performed in 32 wells in August 2000. The remaining 18 wells could not be used for monitoring purpose due to lack of monitoring provision therein. The present monitoring works are being carried out by the GWRDB. The 32 monitoring tube wells are located in the three groundwater districts are Northern, Central and Southern district (as discussed in Figure 3 below). The distributions of the wells are listed as follows: (a) Northern Ground Water District: BB-1, DK-1, DK-2, DK-3b, DK-8, BB-6a, GK-5, GK-4, MH-6a, M-5, M-8, M-11, M-12, M-13 and M-14 and DK-3b; (b) Central Ground Water District: Bal-1a, G-17, G-16, G-13, H-26, H-17, I-26, P-15, M-6, M-7, and BH-LK; (c) Southern Ground Water District: PH-1, P-15, P-7, G-2a, G-48, M-1 and M-4.

Data from the aforementioned districts were used to assess the current trend in groundwater extraction and to explore possibilities of using MSWP water for groundwater recharge activities. It is noted that due to difficulties in acquiring the data, and large data gaps, only four wells (M-1, M-4, M-11 and M-14) were analyzed to assess the long-term trends (from 2001 to 2014) for the Kathmandu Valley.

Groundwater aquifer characteristics review

Over the past years, there have been various studies regarding the geologic system of the Kathmandu Valley, with particular emphasis on the groundwater aquifer characteristics. In this section, an attempt is being made to summarize key findings from various studies in the region. Regarding the overall

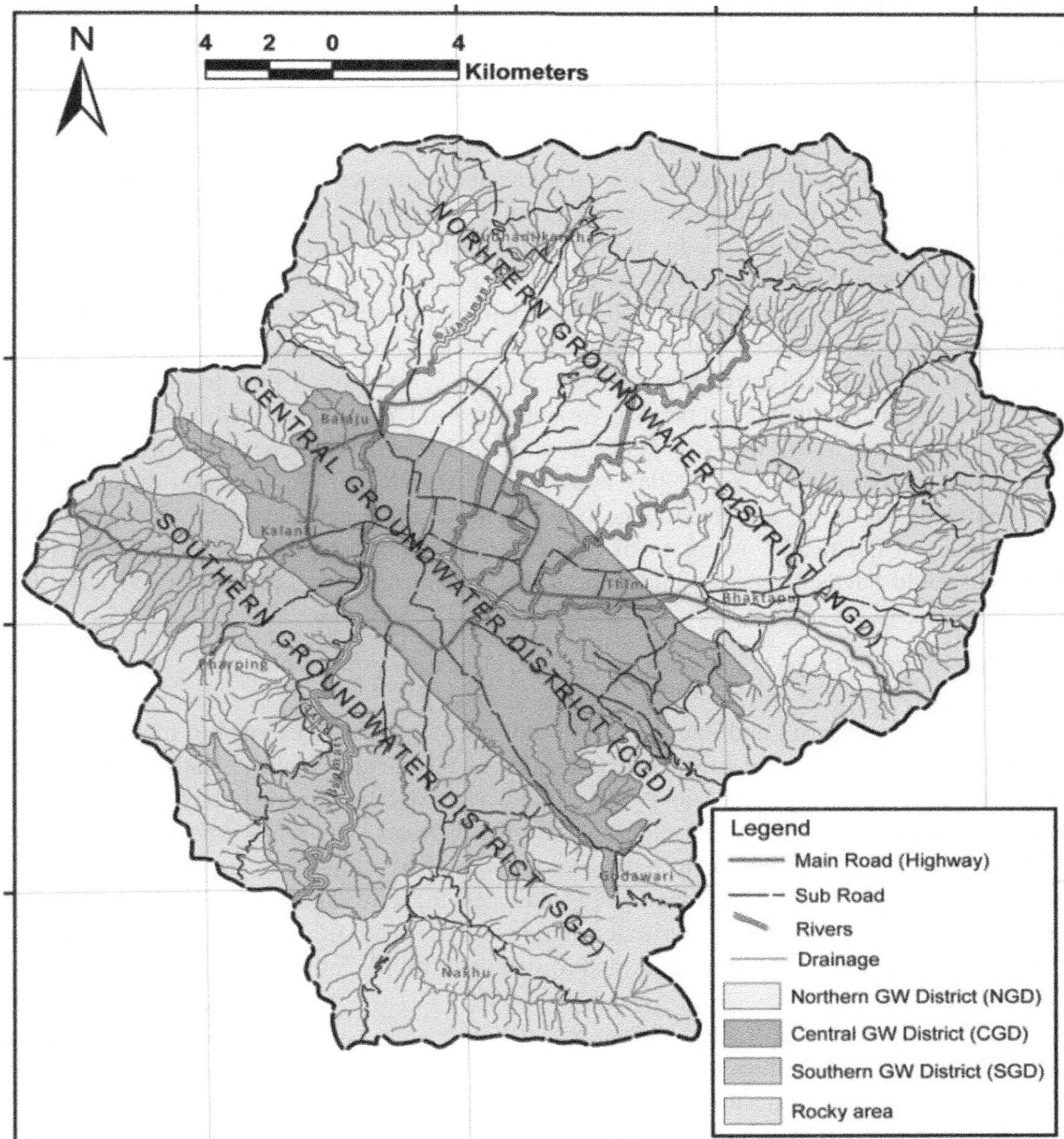

Fig. 3. Groundwater divisions (known as groundwater districts) of Kathmandu Valley based on JICA (1990a, 1990b, 1990c).

stratigraphy for the Kathmandu Valley, it is estimated that fluvio-lacustrine geological deposits are present above the bedrock up to a thickness of 600 m. Above this, the Kathmandu Valley floor has three prominent features: (1) a thick clay layer on the surface, (2) a deep confined aquifer which is (3) confined by basement rocks at the bottom. It is noted that in many regions, the thick clay layer separates the confined aquifer from the unconfined aquifer, thus acting as an aquitard. The clay bed is noted to attain a maximum thickness of 200 m in the south, while less thick on the northern side (DoI, 1998; Kharel *et al.*, 1998). In another study, Gautam & Rao (1991) classified Kathmandu Valley into four groundwater zones: (1) unconfined aquifer zone, (2) and (3) two confined aquifer zones, and (4) no groundwater zone. JICA (1990a) also classified the groundwater divisions into three distinct groundwater districts as shown in Figure 3.

On the other hand, estimating the specific aquifer properties of the Kathmandu Valley has been challenging and therefore, different methods were used by various studies. As a result, research works in the past have brought up various estimations on groundwater recharge volume. For example, Binnie & Partners (1989) estimated a groundwater recharge of 30–40 MLD, while JICA (1990a) calculated 27 MLD, based on simulation model. In another study, Gautam & Rao (1991) estimated 12.63 MLD of groundwater recharge. Kansakar (1996) calculated the groundwater balance of −12.6 MCM (deficit) by estimating rechargeable groundwater resources (dynamic groundwater reserve) of 5.5 MCM and abstraction for drinking purposes and industrial uses of 13.87 MCM and 4.23 MCM, respectively. Stanley *et al.* (1994) estimated the total sustainable withdrawal of groundwater from the valley's aquifers to be 26.3 MLD. Another example for variations would include the estimation of the natural recharge of groundwater in the valley: ranging from 30,000 to 40,000 m^3 per day (Binnie & Partners, 1989), to 15,000 m^3 per day (JICA, 1990a), and about 13,000 m^3 per day (Gautam & Rao, 1991).

It is to be noted that all the aforementioned research needs to be updated, as the dominant land-use pattern has changed drastically owing to the increase in urbanization activities. Also, the recharge characteristics would have changed owing to the ongoing climate change patterns (GWRDB, 2012).

Melamchi Water Supply Project

The MWSP is an inter-basin and inter-sectoral water transfer project designed to meet the long-term (over 30–40 years) water demand of three major cities of Kathmandu Valley (Kathmandu, Patan and Bhaktapur), located in the Bagmati basin, by diverting water from the Melamchi River located upstream of the adjoining Indrāvati basin, which is part of the Koshi River. The major transmission line of the MWSP project is shown in Figure 4 (in red), while a map of existing water supply systems is shown in Figure 5. In Figure 5, the individual color indicates the coverage of the distribution network which is fed either by surface or groundwater supplies. The water from Melamchi will be collected to the Sundarijal/Manhakal chaur system and distributed to the other systems across the valley (MWSDB, 1998).

The project, under the infrastructure development component, includes physical intake and river diversion structures, a 26 km long tunnel, 25 km access road to the tunnel sites, 22 km of approach road, 15 km of main access road to the project intake site, a water treatment plant with a capacity of 170 MLD in Kathmandu, bulk distribution systems (transmission pipe lines and reservoirs in the Kathmandu Valley), improvement of the city water distribution network and its wastewater system, including the construction of wastewater treatment plants (MWSDB, 1998).

Estimation of economic benefits

Water supply rates from the region during wet and dry season will be estimated. In addition, the price to purchase groundwater will be estimated. Since the aim of this study is to assess the increase in recharge along already existing infrastructure, there is no direct costs to the system, other than channeling excess water from the main MWSP channel to the recharge infrastructures via portable pipes or

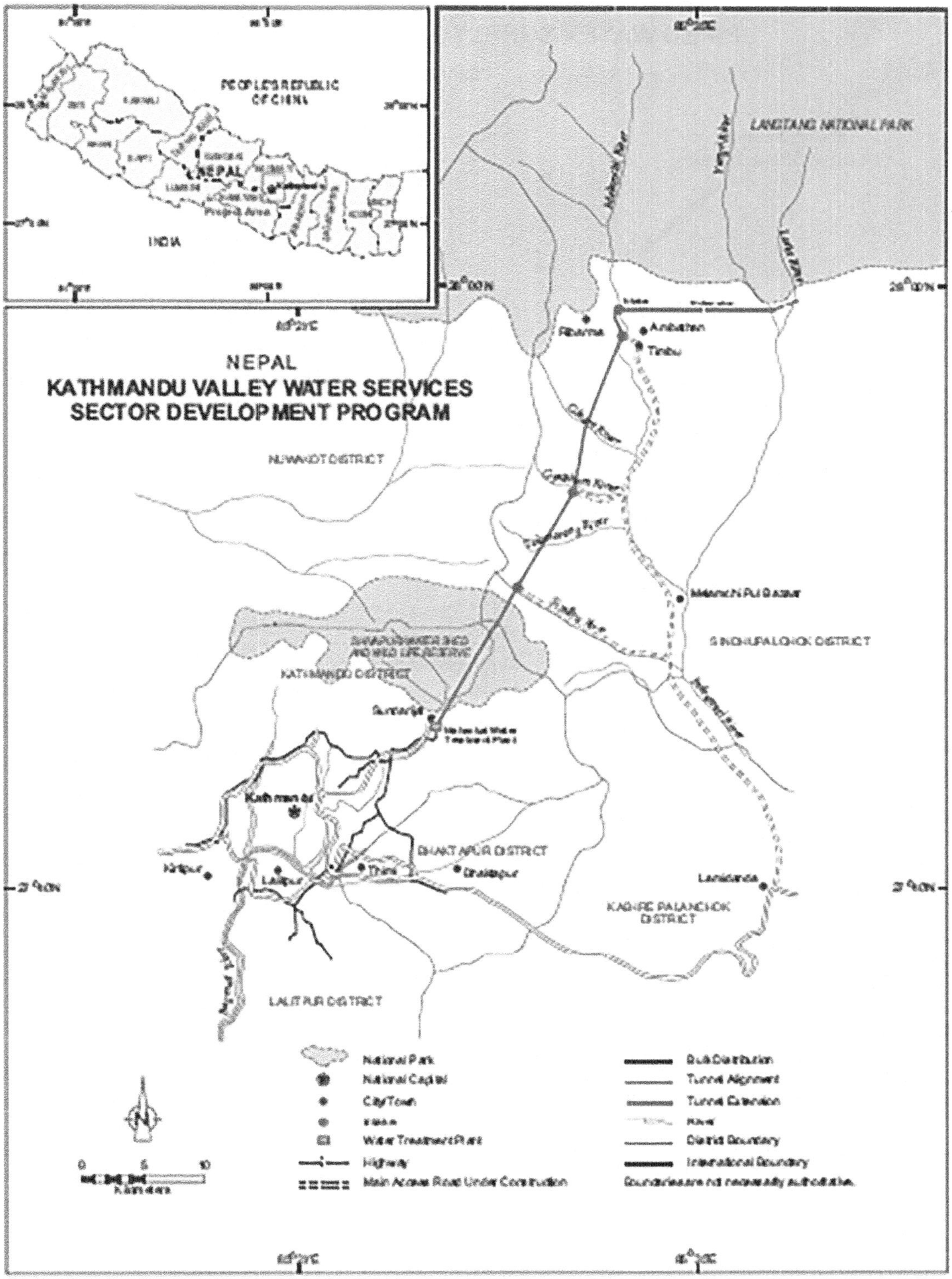

Fig. 4. Kathmandu Valley in Nepal, with the proposed Melamchi water supply scheme (*Source*: MWSDB, 1998). Please refer to the online version of this paper to see this figure in color: http://dx.doi.org/10.2166/wp.2019.080.

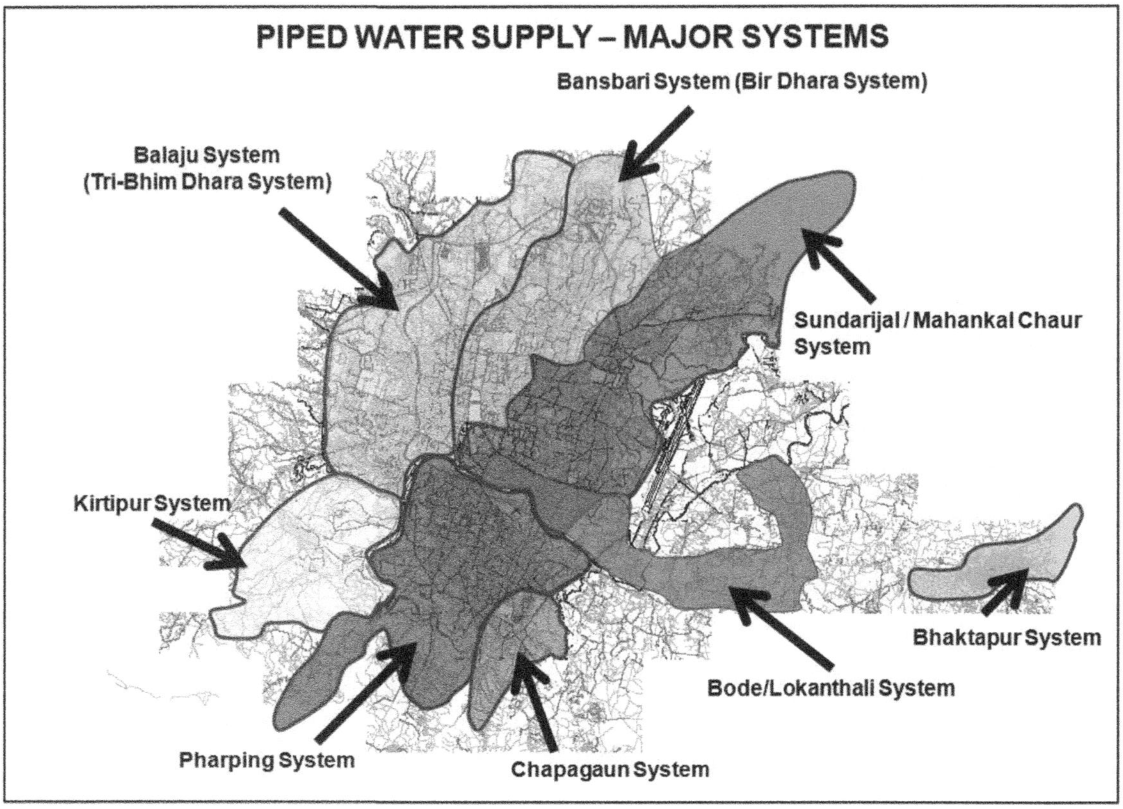

Fig. 5. Existing piped-water major supply systems which will be supplemented under the MWSP scheme for the Kathmandu Valley (*Source*: MWSDB, 1998). Please refer to the online version of this paper to see this figure in color: http://dx.doi.org/10.2166/wp.2019.080.

earthen canals. Such activities can be assumed to be carried out by local organizations (e.g. District or VDC offices) as the benefits are more localized than at large scale.

In order to assess the economic benefits in using MWSP water to recharge groundwater resources, it is necessary to understand the current tariffs for water, both in the government sector and private sector. The existing cost (NRs = Nepal Rupees) of KUKL water supply and private water supply is shown in Table 2.

On the private market, the cost of a 6,000 liters tanker supply of water is generally NRs. 1,200 (NRs 0.2 /liter), while a 12,000 liters tanker costs NRs. 2,000 (NRs 0.17 /liter). Therefore, on average, NRs 0.183 per liter on average.

Results

Groundwater balance in the Kathmandu Valley

As discussed in the earlier sections, groundwater is one of the major water resources to ease the increasing water supply stress for the Kathmandu Valley, which is undergoing rapid urbanization.

Table 2. Water tariffs under the KUKL system and private system for the Kathmandu Valley.

KUKL water supply tariff[a]

Pipe size (Inch)	Maximum limit of use (Liter)	Connection with meter installation: Tariff (NRs)	Connection with meter installation: Additional cost per 1,000 liter beyond maximum limit of use (NRs)	Connection without meter (NRs)
1/2″	10,000	100	32	785
3/4″	27,000	1,910	71	4,595
1″	56,000	3,960	71	9,540
1.5″	155,000	10,950	71	26,280
2″	320,000	22,600	71	54,255
3″	881,000	62,240	71	149,415
4″	1,810,000	17,865	71	306,880

Private water supply tariff[b]

Liters	Tariff (NRs)
6,000	1,200
12,000	2,000

[a]*Source*: http://www.kathmanduwater.org/notice/Tariff%20RateFinal-2070%20Srawon.pdf.
[b]*Source*: Field survey.
NRs = Nepal Rupees (1 USD = 107.2 NRs).

The current study aimed to understand the groundwater balance for the Kathmandu Valley, and hence summarized results from the GWRDB (2012) report. The groundwater balance for the Kathmandu Valley is computed as follows (GWRDB, 2012):

$$\Delta GW = \text{Groundwater Inflow} - \text{Groundwater Outflow} \tag{1}$$

$$\Delta GW = (P + Ir + Sr) - (ET + GD + I + GL + Q) \tag{2}$$

$$GD = (De + Sh + St) \tag{3}$$

where ΔGW is the change in groundwater storage, P, Ir and Sr are groundwater recharge due to rainfall (precipitation), irrigation return flows and supply return flows, respectively, ET is evapotranspiration losses, I is the irrigation water demand, GL is the groundwater loss to the aquifer and Q is the baseflow, GD is the total groundwater draft, which is the sum of draft from deep aquifer (De), shallow aquifer (Sh) and stone spouts (St). All volumes are in MCM.

Computation of the major groundwater balance components according to GWRDB (2012) is discussed in the following sections.

Groundwater recharge and groundwater basin area. According to the reports by the Japan International Cooperation Agency (JICA, 1990a, 1990b, 1990c), the Kathmandu Valley watershed is divided into three parts: Northern, Central and Southern (Figure 3). Out of the total area of the northern part (157 km^2), 59 km^2 is the area available for groundwater recharge. The total area of the central part is 114 km^2, however only 6 km^2 is suitable for groundwater recharge, and of the total area of 55 km^2 in the southern part, groundwater recharge area is 21 km^2. Hence total recharge area of the Kathmandu

Valley is around 86 km^2 (Dixit & Upadhya (2005). In another study by Pandey *et al.* (2010), the aquifer area for the Kathmandu Valley was quantified using GIS layers, giving a net area of 241 km^2. GWRDB (2012) estimates that the groundwater basin of the Kathmandu Valley covers 327 km^2, which is approximately 50% of the total watershed area (650 km^2). For this current study, the latest groundwater basin area estimation by GWRDB (2012) was used (i.e. 327 km^2).

Groundwater recharge from precipitation (P). GWRDB (2012) used precipitation data from the Department of Hydrology and Meteorology (DHM) and estimated that the 34-year average annual rainfall for the Kathmandu Valley is 1,784 mm. With a valley area of 650 km^2, the amount of groundwater recharge from rainfall was estimated to be 1,159.6 MCM per year.

Groundwater recharge from irrigation return flows (Ir). The Department of Irrigation (DoI) estimates that the total irrigated command area in the Kathmandu Valley is 9,120 ha (DoI, 2007). Based on the agricultural activity in the Kathmandu Valley, the GWRDB (2012) estimated that 23.61 MCM per annum was consumed by the crops out of which the return flows due to irrigation water was computed to be 20% (i.e. 4.7 MCM per annum).

Groundwater recharge from transmission losses (Sr). The GWRDB (2012) estimated that 35–40% of the total water supply of KUKL (66.43 MCM) is lost to the groundwater aquifer (26.6 MCM) due to transmission losses, especially due to leakage in the transmission pipes.

Total groundwater draft (GD)

Groundwater extraction from deep aquifer (*De*): GWRDB (2012), from their well inventory study in Kathmandu Valley, indicate that there are 759 deep tube wells in the region, with a total groundwater extraction by private (hotels, industries, companies and housing companies) and community tube wells (organizations, offices, etc.) in the order of 31.15 MCM. In addition, using the water budget report by the Kathmandu Valley Water Supply Management Board (KVWSM, 2010), the annual groundwater extraction by NWSC and KUKL wells in Kathmandu Valley (used for water supply in the region) was estimated to be 38.29 MCM. This shows that the annual extraction of deep aquifer groundwater from the Kathmandu Valley was 69.4 MCM in 2012 (Table 3).

Groundwater extraction from shallow aquifer (*Se*): A huge gap between demand and municipal water supply still prevails in the region, which has given rise to the private water sector. These private players satisfied most of the water demand for the Kathmandu Valley by sourcing water mainly from the groundwater sources such as tube wells and natural spring sources found in and around the Kathmandu Valley region. Water is transported from these sources by vehicles ranging from 2,000 liters (using pickup trucks) to 12,000 liters (using water tankers). Since there are good profit margins in this business, there has been a sharp increase in the number of private water suppliers and private tube wells tapping in to the local aquifers. In a recent study on the water market in the valley, Shrestha & Shukla (2010) indicate that on average 25.5 and 8.5 MLD of water were sold in the market during dry and wet seasons, respectively, during the year 2009. It was also noted that the shallow aquifer was used for extraction of this water. In order to close the groundwater balance, GWRDB (2012) had to estimate the volume of water supplied by the private sector. However, since there is a paucity of data on private water supply, GWRDB (2012) assumed that 50% of the supply gap was supplied by the private sector. For

Table 3. Water balance computation for Kathmandu Valley (*Source*: GWRDB, 2012).

Sl. No.	Water balance component	Water volume (MCM)
Input to the system		
1	Precipitation (*P*)	1,159.6
2	Irrigation return flow (*Ir*)	4.7
3	Water supply return flow (*Sr*)	26.6
Output from the system		
1	Evapotranspiration (*ET*)	521.3
2	Groundwater Draft (*GD*)	
	a. Deep Aquifer (*De*)	69.4
	b. Shallow Aquifer (*Se*)	58.8
	c. Stone Spout (*St*)	2.1
3	Irrigation Water (*I*)	23.6
4	Loss to Deep Aquifer (*GL*)	5.7
5	Baseflow (*Q*)	529.4
Groundwater storage change		−19.4

the year 2012, the supply gap was 61.32 MCM and hence GWRDB (2012) assumed that 50% of this deficit was met by groundwater from shallow aquifer, totaling 30.66 MCM of water extraction from shallow aquifers. In addition to this, KUKL also used shallow wells to supply 28.44 MCM of water, thus bringing the total water extracted in the shallow aquifer to 58.8 MCM (Table 3).

Groundwater extraction from stone spouts (*St*): It is noted that stone spouts, traditional water supply schemes, still supply the water needs of the valley citizens in regions where KUKL water supply is limited. GWRDB (2012) reports that 103, 43 and 78 stone spouts are located in Kathmandu, Lalitpur and Bhaktapur, respectively, thus totaling 224 stone spouts for the Kathmandu Valley. In Kathmandu most of the stone spouts have dried up, while 37 spouts are perennial in Lalitpur and only 27 are currently operational in Bhaktapur. From the district reports on spout discharge (ICON *et al.*, 2002), GWRDB (2012) estimated that the average discharge from these stone spouts is 2.1 MCM annually.

Groundwater loss to deep aquifer (*GL*): Studies in the past have utilised various calculations on the volume of groundwater loss to deep aquifer in the Kathmandu Valley. It is noted that recent scientific studies on this topic have been limited. For the purpose of this study the critical recharge is calculated on the basis of latest report published by JICA in 1993 (JICA, 1993). According to this report, the annual groundwater loss to deep aquifer recharge is 5.7 MCM per year.

Baseflow discharge and evapotranspiration (*Q* and *ET*): GWRDB (2012), using discharge data from DHM (Khokana Station installed by DHM, station Index No. 550.05), estimated the average annual baseflow discharge of the Baghmati River to be 529.44 MCM per annum. Using evapotranspiration (*ET*) data from DHM, the annual average *ET* was estimated to be 521.3 MCM.

Using the results from the aforementioned subsections, the groundwater balance was estimated by GWRDB (2012) as shown in Table 3, with an annual deficit of 19.4 MCM.

From the aforementioned results, the GWRDB (2012) indicates a net water deficit of 19.4 MCM for the 2012 year in the Kathmandu Valley. These results were in agreement with previous studies, for example, Kansakar in 1996 estimated a deficit of 12.6 MCM, while HYM Consult (1997) estimated a deficit of 22.71 MCM in year 1993. The aforementioned results all collectively show that the aquifer in the Kathmandu Valley has been constantly under stress due to over-extraction activities. Such activities may not be sustainable in the long term, and could cause high socio-economic stress (Chinnasamy & Agoramoorthy 2015, Chinnasamy & Prathapar 2016).

Long-term groundwater level trend

The previous section indicated that there is over-extraction of groundwater in the basin. The current study analyzed long-term groundwater level trends to assess the groundwater depletion in the regions that supply water for Kathmandu. Long-term groundwater level data from years 2001 to 2013 in four wells were used for this purpose. For the purpose of orientation, within the Kathmandu Valley, the observation well M1 (Figure 6(a)) is located in Kirtipur at the south, the observation well M4 (Figure 6(b)) is located in Lubhu at the south east, the observation well M11 (Figure 6(c)) is located in Danchhi and the observation well M14 (Figure 6(d)) is located in Mulpani at the north. It is noted that the Mulpani region lies in the recharge area of deep aquifer of the valley. In addition, it is noted that for most cases the water level data for the years 2005–2007 and 2009 were not available.

Long-term groundwater trend results indicate that all wells have a net decline in groundwater levels (Figure 6(a)–6(d)). The observation well in Lubhu (Lalitpur) recorded the greatest decline of 1.27 m per year. The observation wells at Mulpani and Kirtipur recorded groundwater decline of 0.46 and 0.34 m per year, respectively. The observation well in Danchhi recorded the least decline with a rate of 0.06 m per year.

It is also noted that, records from the well fields of the NWSC in the deep aquifer have shown a drawdown of the surface by 15–20 m since the construction of the wells in 1984/85 indicating substantial over-exploitation. Both static and pumping water level have been depleted in most parts of the Kathmandu Valley. This trend is verified by the water level hydrographs from GWRDB monitoring wells across the valley. Data from 2001 to 2014, which indicate a decline in water level by 2 m in Kirtipur, 10 m in Lubhu, 3 m in Mulpani and 10 m in Koteshwor.

The overall analysis shows that the decline in water level is not occurring in some isolated part of the valley but in the whole of the Kathmandu Valley. In addition, it is reported that there are about 103, 43 and 78 numbers of stone spouts in Kathmandu, Lalitpur and Bhaktapur, respectively. In Kathmandu most of the stone spouts are dried up. Out of 78 stone spouts in Bhaktapur, only 27 are operational. The drying of these spouts and the lowering of water levels in the observation wells indicate that the decline of water level is taking place in both shallow and deep aquifers.

Groundwater storage volume assessment

As performed in India in a similar geological setting, the water table fluctuation method was used to assess the groundwater storage volume in the basin (CGWB, 2015). According to this method, the groundwater storage volume (*GW*) due to change (Δ) in groundwater level (*GL*) is assessed as follows (Equation (4)):

$$GW = \Delta GL * S_y * A \tag{4}$$

where GW is in cubic meters (m^3), ΔGL is in meters (m), S_y is the specific yield – a hydrologic parameter of the aquifer (unitless) and A is the area of the aquifer in meters squared (m^2). Pandey *et al.* (2010)

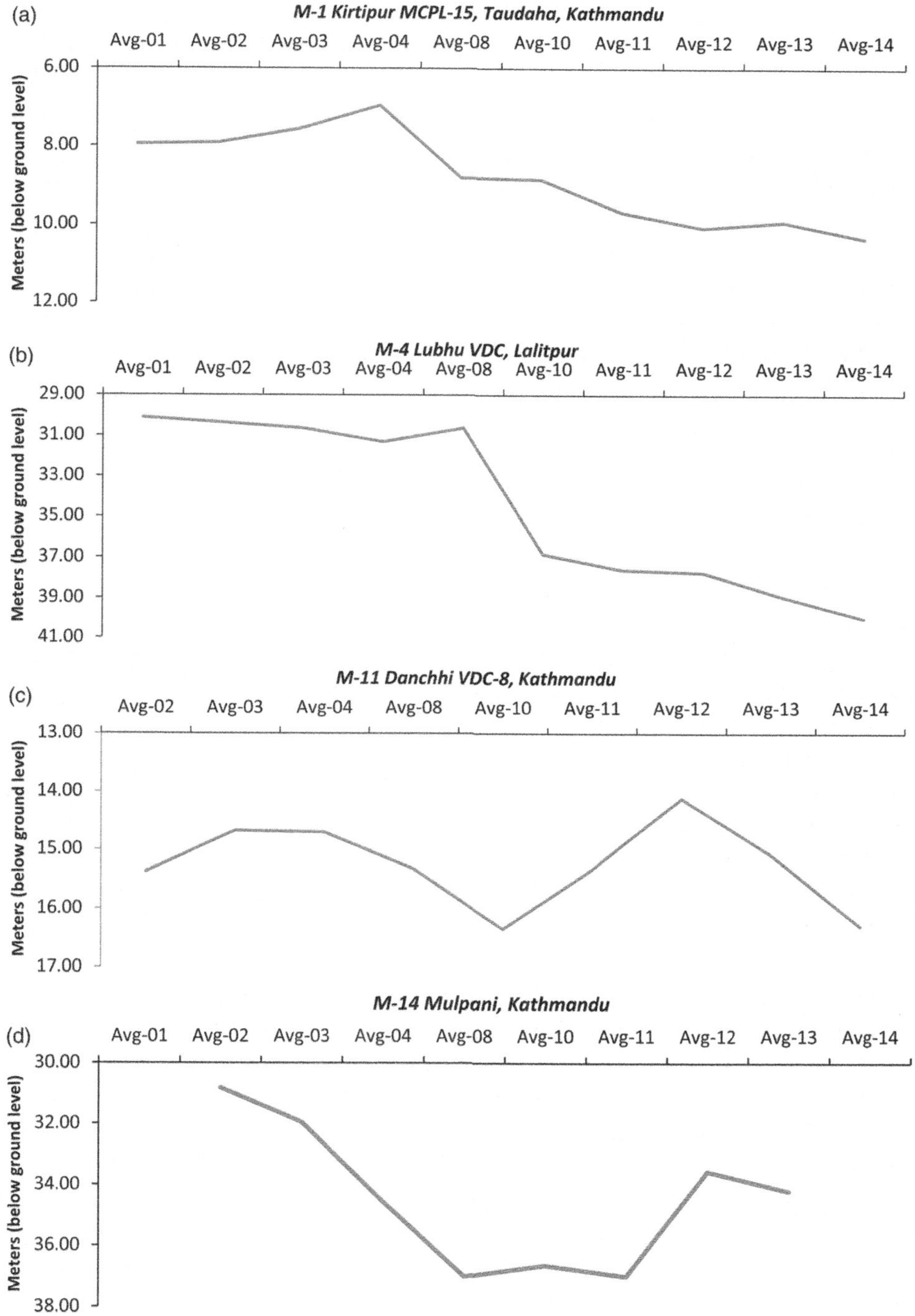

Fig. 6. Long-term groundwater trends for selected observation wells in the Kathmandu Valley: (a) Kirtipur, (b) Lubhu, (c) Danchhi and (d) Mulpani.

reported that the specific yield of the shallow aquifers for the Kathmandu Valley was 0.2, while GWRDB (2012) estimated the area of the aquifer to be 327 km^2 with an average ΔGL of 3.76 m over the study period from 2001 to 2014 (using the four observed wells). Using these values in Equation (4), it was estimated that the current depletion in the groundwater aquifers of the Kathmandu Valley can store an average of 246 MCM. Since the unmet demand in the valley is 168 MLD (Table 1), this groundwater storage volume can easily aid in easing the water stress for the Kathmandu Valley. In addition, groundwater storage can be recharged during the wet season wherein excess water can be diverted to recharge structures, while water can be pumped from this source during the dry season. It is noted that all the recharge water need not come from the MWSP scheme alone as there is enough excess water that leaves this basin during the wet season (Chinnasamy *et al.*, 2015; Chinnasamy, 2016b). It can be noted that the 246 MCM volume was based on the groundwater depletion over the study period (2001–2014) and hence the annual unmet demand of 61 MCM (equivalent of 168 MLD) can be eased even at annual timescales if the groundwater resources are recharge sustainably.

Economic benefits

The current MWSP scheme is reported to supply an additional 170 MLD for the Kathmandu Valley. However, this supply volume is still less than the unmet demand in the region when the aforementioned groundwater storage is not taken into account (Table 1). In the event that excess water is being diverted to the groundwater recharge structures in the Kathmandu Valley, it is assumed that all water harvesting structures will be filled up with water during excess flow regimes. This is because the Kathmandu Valley basin is not a water deficit basin, especially during wet seasons, but a basin suffering from unmet water demands due to lack of water management initiatives (Chinnasamy *et al.*, 2015). There is tremendous flow in the basin during wet seasons, while the dry seasons have low flow. Therefore, while all the water demands are met in the basin during wet seasons, there are huge unmet demands during the dry seasons. According to Chinnasamy *et al.* (2015), the Koshi basin, which contains the Kathmandu Valley, generates 8,262 MCM of surface runoff water annually with 7,368 MCM during the wet season (90% of total annual flow). The study also noted that the current unmet demand was only 660 MCM, and hence storing/diverting excess water would be an economically profitable method to ease water stress.

If proper management plans are laid, wherein the MWSP supply scheme also channels excess water to the recharge zones of the basin, the current groundwater storage depletion can be filled. Under this trend, a total volume of 246 MCM of water can be stored in the groundwater reserves in the valley during the wet season. This water can then be pumped and supplied to the public at a fraction of the cost that tankers charge during dry season. As per the previous sections, during the dry season, on average, water is sold by private agencies at 0.17 NRs per liter, and 0.009 NRs by the government agencies (Table 2). Even securing this water is very tough due to long wait lines, and fuel shortages for supply vehicles. Therefore, assuming that 246 MCM can be stored in the groundwater reserves and that the government agencies ration this water under their current tariff, groundwater storage can save up to 2.2 billion NRs for the valley during the dry season, and also regulate the supply of water throughout the dry season. In addition to the costs, the time that women spend collecting water can be greatly reduced, thus improving their quality of life, and the excess water can be used for better sanitation facilities. With proper management plans and institutional policy amendments, the excess water can be used to recharge the groundwater storage stores in the Kathmandu Valley.

Discussion

Earlier studies indicated that the groundwater extraction came into existence in the Kathmandu Valley from 1970s, with initial extraction less than 0.04 MCM per year. The impact of this extraction was minimal on the groundwater status, but by the 1980s the extraction was 3.51 and 12.3 MCM per year during 1982 and 1987, respectively. This extraction caused visible impacts on the groundwater levels with a slight decline in groundwater levels. From the 1990s haphazard pumping caused significant impact on groundwater levels. The pumping rates were 18.7 and 21.26 MCM in 1993 and 1999, respectively (Kharel *et al.*, 1998; Metcalf & Eddy, 2000; Pandey *et al.*, 2010).

Comparing the groundwater levels between 2013 and 1984, a drastic decline in groundwater level in the deep aquifers was inferred, some aquifer depletion rates ranging from 1 to 4 m per year. It was also inferred that most of the regions impacted by high groundwater level decline were in or near regions with high groundwater abstraction rates, especially the northern–northeastern parts of the Kathmandu Valley. In other studies, failure of deep wells, dug wells and dhunge dharas (traditional stone water spouts) due to the over-extraction have been reported across the valley. Systems dependent on shallow aquifers, namely dug wells, dhunge dharas and ponds were failing mostly due to inadequate water for recharge. It is noted that wells in the older parts of the town, e.g. Patan sub-metropolis, have survived as the rate of increase in urbanization activities is less (Shrestha, 2009; GWRDB, 2012). Another common report is the drying up of newly drilled bore and deep wells, which gave water for a short period from stagnant storage, but eventually failed due to falling groundwater levels.

It is noted by many studies that the increasing water demand for the Kathmandu Valley population cannot be solely satisfied by current or proposed water supply schemes by the federal agencies and from municipal corporations (Cresswell *et al.*, 2001; Dixit & Upadhya, 2005; Pattanayak *et al.*, 2005; Shrestha, 2012). The inadequate and inefficient supply systems of municipal corporations have led the population at large to supplement their water supply by tapping into traditional water sources, i.e., stone spouts (Shrestha, 2009; Shrestha *et al.*, 2013). Unfortunately many stone spouts are now converted into temporary waste dumps that require proper rehabilitation.

Future directions

First and foremost, as noted in the review sections, there are many inconsistencies in assessing the properties of the aquifers and data on water demand and supply. Therefore, future studies should aim to collect and archive quality data that can be used to accurately estimate water demand and supply in the region. Such initiatives should be undertaken by federal agencies that have a large stake in water supply and management in the valley.

Chapagain (2014) notes that the unplanned and unregulated urbanization in the Kathmandu Valley has been the primary reason for inadequate water supply in the region. As a result, Chapagain (2014) urges the federal agencies to consider water supply schemes while undertaking urbanization plans. However there exist many challenges in increasing groundwater recharge by using present water recharge structures, mostly related to anthropogenic activities (Dixit & Upadhya, 2005). Results from this study will aid in sensitizing ongoing groundwater issues in the Kathmandu Valley to the government officials and general public. With such knowledge, the strategic distribution of Melamchi water can aid in easing water stress and also aid indirectly by increasing groundwater reserves which will ease dry season water stress.

With the MWSP it is assumed that better channeling of water will occur due to improved infrastructure in water transmission lines. Therefore, there is a school of thought that with the MWSP water supply scheme, the water leakage during transmission will be lower than the current rate of 35–40%, and as a result the amount of water indirectly recharged to the aquifer would be less under the MWSP scheme. While the new MWSP infrastructure would indeed cut transmission losses, the possibility for channelizing excess water from MWSP and seasonal rainfall would be high. As a result, there will be more water to directly induce more recharge to the aquifers. Future studies should therefore attempt to quantify any change in groundwater recharge due to the MWSP scheme. Groundwater simulation models, e.g. MODFLOW, can be used to use these estimated changes in groundwater recharge to assess the distribution of groundwater volume across the basin (Chinnasamy, 2012; Chinnasamy & Hubbart, 2014a, 2014b).

Udmale *et al.* (2016) estimated the water deficit (demand – supply) in a recent study and found that even with the completion of the MWSP, there would be a deficit of 102 MLD for the Kathmandu Valley. In addition, they note that if the MWSP is completed in time, along with other water harvesting methods, especially treatment and distribution infrastructure, the water deficit can be eliminated by 2025.

Future studies should concentrate more on the fact that the MWSP alone cannot solve the water crisis for Kathmandu Valley (Laia *et al.*, 2013), but a more proactive hydraulic and water accounting models, such as Water Evaluation and Planning (WEAP) (Chinnasamy *et al.*, 2015) can be used to assess optimum conditions to release water to recharge structures within the MWSP plan, thereby not affecting downstream water users. In addition, scientifically validated decision support tools (DSTs) should be developed specifically for the region to augment sustainable and conjunctive use of surface and groundwater resources for the domestic supply of the ever-growing Kathmandu Valley. Furthermore, various scenarios should be analyzed using the developed DSTs, to increase the efficiency of water conservation and management efforts. Conjunctive use of surface and groundwater resources, increasing groundwater recharge rates by using MWSP and introduction of rainwater harvesting structures are key to having no water deficit by 2023.

Conclusion

Preliminary results from the current study indicate that the groundwater of the valley is being consumed at unsustainable rates, thus leading to widespread groundwater depletion. Due to this depletion, it was estimated that a groundwater storage of 246 MCM has been created. If this storage is recharged by excess runoff in the wet season, the stored water can easily satisfy the ongoing domestic water deficit of 178 MCM per annum. The timely implementation of the MWSP can aid in easing the ongoing water stress and can aid in reversing the damage caused to groundwater storage due to the delay of the project if appropriate infrastructure are planned that can harvest excess surface runoff during the monsoon months. Therefore, in the long run, MWSP water can supply water and recharge groundwater during the monsoon period, and groundwater resources can ease supply stress during the dry seasons, thus improving quality of life and socio-economic status in Kathmandu. Such a conjunctive use scenario will aid in improving the livelihood for the citizens of Kathmandu Valley.

Acknowledgements

The authors gratefully acknowledges the financial support provided by the Consortium of International Agricultural Research Centers (CGIAR) Research Program on Water, Land and Ecosystems (WLE) and the Ground Water Resources Development Board, Ministry of Irrigation, Government of Nepal. The author thanks the Ground Water Resources Development Board for the various reports related to the groundwater of Kathmandu Valley. The authors would also like to thank the anonymous reviewers for their constructive reviews.

References

ADB (Asian Development Bank) (2000). *Report and Recommendation of the President to the Board of on A Proposed Loan to the Kingdom of Nepal for the Melamchi Water Supply Project, November 2000*. RRP: NEP 31624. The Asian Development Bank, Kathmandu, Nepal.

ADB (Asian Development Bank) (2010). *Kathmandu Valley Water Supply and Waste Water System Improvement: TA 4893-NEP: Project Feasibility Final Report*. The Asian Development Bank, Kathmandu, Nepal.

Agoramoorthy, G., Chaudhary, S., Chinnasamy, P. & Hsu, M. J. (2016). Harvesting river water through small dams promote positive environmental impact. *Environmental Monitoring and Assessment 188*, 645.

Amatya, P. (2006). *Traditional Stone Spout Enumeration, Mapping and Its Water Quality (Five Municipal Areas of Kathmandu Valley)*. NGO Forum, WaterAid.

Becker-Ritterspach, R. O. A. (1995). *Water Conduits in the Kathmandu Valley*. Munshiram Manoharlal Publishers, New Delhi.

Binnie and Partners (1973a). *Master plan for water supply and sewerage of the greater Kathmandu and Bhaktapur*, Vol. 1 & 2, Water Supply, WHO UNDP (special fund) Project Nepal 0025, Water Supply and Sewerage Corporation, Nepal. WHO, Geneva.

Binnie and Partners (1973b). *Kathmandu Water Supply and Sewerage Scheme*. Appendix 4.5, Groundwater Investigations. World Health Organization, Project Nepal 0025. Unpublished.

Binnie and Partners (1989). *Water Supply for Kathmandu – Lalitpur From Outside the Valley*. Prefeasibility study, Final Report. Environment and Public Health Organization (ENPHO). Unpublished.

Central Bureau of Statistics (2011). *National Population and Housing Census 2011 (Village Development Committee/Municipality)*. Government of Nepal, Kathmandu, Nepal.

CGWB (Central Ground Water Board) (2015). *Year Book 2015*. Central Ground Water Board, Ministry of Water Resources, Government of India, Delhi, India.

Chapagain, D. P. (2014). Urban water supply sector reform in Kathmandu Valley. *Journal of the Institute of Engineering 9*, 130–141.

Chinnasamy, P. (2012). *Measuring and Modeling Water and Nutrient Flux Between A mid-Missouri Stream and Forested Riparian Zone in the Central US. Doctoral dissertation*, University of Missouri, Columbia, USA.

Chinnasamy, P. (2016a). Depleting groundwater – an opportunity for flood storage? A case study from part of the Ganges River basin, India. *Hydrology Research 48*, 431–441.

Chinnasamy, P. (2016b). Inference of basin flood potential using nonlinear hysteresis effect of basin water storage: case study of the Koshi basin. *Hydrology Research 48*, 1554–1565.

Chinnasamy, P. & Agoramoorthy, G. (2015). Groundwater storage and depletion trends in Tamil Nadu State, India. *Water Resources Management 29*(7), 2139–2152.

Chinnasamy, P. & Agoramoorthy, G. (2016). India's groundwater storage trends influenced by tube well intensification. *Groundwater 54*, 727–732.

Chinnasamy, P. & Hubbart, J. A. (2014a). Measuring and modeling shallow groundwater flow between a semi-karst border stream and Ozark forested riparian zone in the central USA. *Journal of Scientific Research and Reports 3*, 844–865.

Chinnasamy, P. & Hubbart, J. A. (2014b). Potential of MODFLOW to model hydrological interactions in a semi-karst floodplain of the Ozark border forest in the central US. *Earth Interactions 18*, 1–24.

Chinnasamy, P. & Prathapar, S. A. (2016). *Methods to Investigate the Hydrology of the Himalayan Springs: A Review*. Working Report 169. International Water Management Institute, Colombo, Sri Lanka.

Chinnasamy, P., Bharati, L., Bhattarai, U., Khadka, A., Dahal, V. & Wahid, S. (2015). Impact of planned water resource development on current and future water demand in the Koshi River basin, Nepal. *Water International 40*, 1004–1020.

Chinnasamy, P., Maheshwari, B. & Prathapar, S. A. (2018a). Adaptation of Standardised Precipitation Index for understanding watertable fluctuations and groundwater resilience in hard-rock areas of India. *Environmental Earth Sciences 77*, 562.

Chinnasamy, P., Muthuwatta, L., Eriyagama, N., Pavelic, P. & Lagudu, S. (2018b). Modeling the potential for floodwater recharge to offset groundwater depletion: a case study from the Ramganga basin, India. *Sustainable Water Resources Management 4*, 331–344.

Cresswell, R. G., Bauld, J., Jacobson, G., Khadka, M. S., Jha, M. G., Shrestha, M. P. & Regmi, S. (2001). A first estimate of ground water ages for the deep aquifer of the Kathmandu Basin, Nepal, using the radioisotope chlorine-36. *Ground Water 39*, 449–457.

Dixit, A. & Upadhya, M. (2005). *Augmenting Groundwater in Kathmandu Valley: Challenges and Possibilities*. Nepal Water Conservation Foundation, Kathmandu, Nepal.

DoI (Department of Irrigation) (1998). *Hydrogeological Conditions and Potential Barrier Sediments in the Kathmandu Valley*. Technical Cooperation Project-Environmental Geology, Final report, pp. 17-59, Appendix 3. Department of Irrigation, Government of Nepal, Kathmandu, Nepal.

DoI (Department of Irrigation) (2007). *District Agricultural Statistics*. Department of Irrigation, Government of Nepal, Kathmandu, Nepal.

Gautam, R. & Rao, G. K. (1991). Groundwater resource evaluation of the Kathmandu Valley. *Journal of Nepal Geological Society 7*, 39–48.

GWRDB (Ground Water Resources Development Board) (2012). *Groundwater Balance in the Kathmandu Valley*. Final report, Groundwater Resources Development Board, Department of Irrigation, Government of Nepal, Kathmandu, Nepal, pp. 1–32.

HMG (2004). *Optimizing Water use in Kathmandu Valley (ADB-TA) Project*. Final Report. Ministry of Physical Planning and Works, Kathmandu, Nepal.

HMG and IUCN (1995). *Regulating Growth: Kathmandu Valley, Main Report of National Conservation Strategy Implementation, Project*. IUCN, Kathmandu Valley, Nepal, pp. 47–49

HMG and MoPE (Ministry of Population and Environment) (1999). *Environmental Planning and Management of the Kathmandu Valley*. MoPE, Kathmandu Valley, Nepal, p. 38.

HYM (Hydrology and Meterology) Consult (1997). Final Report on Water Balance Study of Bagmati Basin within Kathmandu Valley, Kathmandu, Nepal.

ICON, UNESCO and RCUWM (2002). *Situation of Traditional Water Spouts in Kathmandu Valley*. UNESCO, New York.

Jadeja, Y., Maheshwari, B., Packham, R., Bohra, H., Purohit, R., Thaker, B., Dillon, P., Oza, S., Dave, S., Soni, P., Dashora, Y., Dashora, R., Shah, T., Gorsya, J., Katara, P., Ward, J., Kookana, R., Singh, P. K., Chinnasamy, P., Goradiya, V., Prathapar, S., Varua, M. & Chew, M. (2018). Managing aquifer recharge and sustaining groundwater use: developing a capacity building program for creating local groundwater champions. *Sustainable Water Resources Management 4*, 317–329.

Jha, P. K. & Shrestha, K. K. (2010). Climate change and urban water supply: adaptive capacity of local government in Kathmandu City, Nepal. *Journal of Forest and Livelihood 11*, 62–81.

Jha, M. G., Khadka, M. S., Shrestha, M. P., Regmi, S., Bauld, J. & Jacobson, G. (1997). *The Assessment of Groundwater Pollution in the Kathmandu Valley*. AGSO/GWRDB, Kathmandu, Nepal, p. 5.

JICA (Japan International Cooperation Agency) (1990a). *Groundwater Management Project in the Kathmandu Valley*. Final Report to Nepal water supply cooperation. JICA, Tokyo.

JICA (Japan International Cooperation Agency) (1990b). *Groundwater Management Project in the Kathmandu Valley*. Supporting Report. JICA, Tokyo (Unpublished).

JICA (Japan International Cooperation Agency) (1990c). *Groundwater Management Project in the Kathmandu Valley*. Data Book. JICA, Tokyo (Unpublished).

JICA (Japan International Cooperation Agency) (1993). *Master Plan Study for Water Resources Development of Upper Karnali and Mahakali River Basins, Kathmandu*. JICA, Tokyo.

Kansakar, S. (1996). *Water Resources Strategy*. Water Energy Commission Secretariat (WECS), Kathmandu, Nepal.

Kathmandu Valley Water Supply Management Board (2006). *Water Supply Management Board Act*. Government of Nepal, Kathmandu, Nepal.

Kharel, B. D., Shrestha, N. R., Khadka, M. S., Singh, V. K., Piya, B., Bhandari, R., Shrestha, M. P., Jha, M. G. & Mustermann, D. (1998). *Hydrogeological Conditions and Potential Barrier Sediments in the Kathmandu Valley*. Final Report, p. 28.

Khatiwada, N. R., Takizawa, S., Tran, T. V. N. & Inoue, M. (2002). Groundwater contamination assessment for sustainable water supply in Kathmandu Valley, Nepal. *Water Science and Technology 46*(9), 147–154.

KUKL (Kathmandu Upatyaka Khanepani Limited) (2012). *Fifth Anniversary, Annual Report 2069 (2012).* KUKL, Kathmandu, Nepal.

KVWSM (Kathmandu Valley Water Supply Management) (2010). *Annual Report 2010.* KVWSM Board, Kathmandu, Nepal.

KVWSWSI (Kathmandu Valley Water Supply and Wastewater System Improvement) (2010). *Kathmandu Valley Water Supply and Wastewater System Improvement Project Feasibility Study.* PPTA 4893-NEP Final Report. ADB and KUKL, Kathmandu, Nepal.

Lacombe, G., Chinnasamy, P. & Nicol, A. (2019). *Review of Climate Change Science, Knowledge and Impacts on Water Resources in South Asia.* Background Paper 1. World Bank, Washington, DC.

Laia, D., March, H. & Saurí, D. (2013). Contesting large-scale water supply projects at both ends of the pipe in Kathmandu and Melamchi Valleys, Nepal. *Geoforum 47*, 22–31.

Metcalf & Eddy Inc. (1999). *Urban Water Supply Reforms in the Kathmandu Valley, Groundwater Monitoring Program in Association with CEMAT Consultant.* Metcalf & Eddy Inc, Wakefield, MA.

Metcalf & Eddy Inc. (2000). *Urban Water Supply Reforms in the Kathmandu Valley (ADB TA Number 2998-NEP), Completion Report, Volume I & II.* Executive summary, Main report & Annex 1–7. Metcalf and Eddy, Inc. with CEMAT Consultants Ltd, 18 Feb 2000. Metcalf & Eddy Inc, Wakefield, MA.

Metcalf & Eddy Inc. (2002). *Urban Water Supply Reforms in the Kathmandu Valley, Ex. Summary and Final Report, ADB TA Number 2998-NEP, 18th Feb, 2002, pp. 1–1.* Metcalf & Eddy Inc, Wakefield, MA.

MoPE, ICIMOD, SACEP, NORAD and UNEP (2001). *State of the Environment, Nepal.* MoPE, Kathmandu, Nepal, pp. 121–122, 132.

MWSDB (1998). *Melamchi Water Supply Project. Information Brochure.* Melamchi Water Supply Development Board (MWSDB), Government of Nepal, Kathmandu, Nepal, p. 2.

Pandey, V. P., Chapagain, S. K. & Kazama, F. (2010). Evaluation of groundwater of Kathmandu Valley. *Environmental Earth Sciences 60*, 1329–1342.

Pattanayak, S., Yang, J., Whittington, D. & Bal Kumar, K. (2005). Coping with unreliable public water supplies: averting expenditures by households in Kathmandu, Nepal. *Water Resources Research 41*, 1–11.

Shrestha, R. R. (2009). *Rainwater Harvesting and Groundwater Recharge for Water Storage in the Kathmandu Valley.* Sustainable Mountain Development, No. 56, International Centre for Integrated Mountain Development (ICIMOD), Winter 2009, p. 28. ICIMOD, Lalitpur, Nepal.

Shrestha, M. N. (2012). Groundwater use in the Kathmandu Valley: an analysis of pre- and post- melamchi scenarios. In: *Kathmandu Valley Groundwater Outlook.* Shrestha, S., Pradhananga, D. & Pandey, V. P. (eds). Asian Institute of Technology (AIT): Klong Luang, Pathumthani, Thailand/The Small Earth Nepal (SEN): Kathmandu, Nepal/Center of Research for Environment Energy and Water (CREEW): Kathmandu, Nepal/International Research Center for River Basin Environment-University of Yamanashi (ICRE-UY), Kofu, Yamanashi, Japan, Ch. 10, pp. 90–99.

Shrestha, P. & Shukla, A. (2010). *Climate Change Impact on River Dynamics of the Bagmati Basin, Kathmandu, Nepal, Final Report.* National Adaption Program of Action (NAPA) to Climate Change Project, Government of Nepal, Kathmandu, Nepal, pp. 5–30.

Shrestha, A., Karki, K., Shukla, A. & Sada, R. (2013). *Groundwater Extraction: Implications on Local Water Security of Peri-Urban, Kathmandu, Nepal.* Paper, (7). Peri Urban Water Security Discussion Paper Series.

Stanley, I. L., Mott, M. L. & East, C. P. L. (1994). *Bagmati Basin Water Management Strategy and Investment Programme: Final Report.* Kathmandu, Nepal: His Majesty's Government, Ministry of Housing and Physical Planning/Tokyo: JICA/New York: The World Bank.

Udmale, P., Ishidaira, H., Thapa, B. R. & Shakya, N. M. (2016). The status of domestic water demand: supply deficit in the Kathmandu Valley, Nepal. *Water 8*, 196.

The structure of water vending markets in Kathmandu, Nepal

Aditi Raina[a], Jane Zhao[b,c,*], Xun Wu[d], Laxman Kunwar[e] and Dale Whittington[a,b,c]

[a]*Lee Kwan Yew School of Public Policy, National University of Singapore, Singapore*

[b]*Department of Environmental Sciences & Engineering, University of North Carolina at Chapel Hill, Chapel Hill, NC, USA*

[c]*Global Development Institute, University of Manchester, Manchester, UK*

**Corresponding author. E-mail: janezhao@live.unc.edu*

[d]*Division of Public Policy, Hong Kong University of Science and Technology, Hong Kong, China*

[e]*Central Department of Population Studies, Tribhuvan University, Kathmandu, Nepal*

Abstract

To date, there has been limited empirical research on the structure of informal water vending markets in developing countries. From fieldwork conducted in Kathmandu in 2014, including a survey of different types of water vendors, household interviews, and in-depth interviews with key informants, we provide a detailed description of the activities of multiple types of water vendors and examine the profitability of tanker truck vendors and water source vendors. We find that several distinctive markets operate along the supply chain between source water and end users. We conclude that a detailed understanding of the different vending activities in which water vendors engage is essential to the design of public sector policy interventions aimed at improving performance in informal water markets.

Keywords: Informal markets; Market structure; Water vending

Introduction

Despite decades of effort by governments and donor agencies to expand piped water connections, water vending continues to play a major role in water supply for urban residents in many parts of Africa (Whittington *et al.*, 1991; Collignon & Vézina, 2000; Zuin *et al.*, 2014), South America (Casey *et al.*, 2006; Wutich *et al.*, 2016), and Asia (Crane, 1994; Conan & Paniagua, 2003). Because of the unreliability and poor service of piped water systems in many countries, even households connected to a piped network often rely heavily upon private vendors as a major source of water supply (Keener *et al.*, 2009; Ahlers *et al.*, 2013). In fact, inviting water vendors into public–private partnerships in the urban water supply and sanitation sector has emerged as a policy option (Opryszko *et al.*, 2009).

While there is no doubt that water vending has greatly improved access to potable water for millions of people, especially the poor, in developing countries, the social, economic, and environmental

doi: 10.2166/wp.2019.181

consequences of reliance on informal water markets need careful examination. Water supplied by informal water markets may pose serious human health risks (Zaroff & Okun, 1984; Kjellén & McGranahan, 2006; Appiah Obeng *et al.*, 2010) and undermine policies to improve social equity (Wutich *et al.*, 2016).

It is often assumed that the presence of many private water vendors, coupled with relatively few entry barriers, would keep such markets competitive (Ahlers *et al.*, 2013). Yet, the high prices charged by private water vendors, ranging from 400 to 100 times the amounts charged under municipal water tariffs (Zaroff & Okun, 1984; Bhatia & Falkenmark, 1993; Crane, 1994; Snell, 1998; Solo, 1999), raise concerns about the presence of 'monopoly rent' in such markets (Whittington *et al.*, 1991). The competitiveness of informal water markets may also be undermined by collusion among providers and by price fixing by water vendors' business associations (Whittington *et al.*, 1991; Crane, 1994; Snell, 1998; Collignon & Vézina, 2000; Ahlers *et al.*, 2013).

There has been limited empirical research on the structure and competitiveness of informal water markets (Ahlers *et al.*, 2013). Researchers face several methodological challenges. First, record keeping and reporting among vendors operating in informal markets are generally poor. Second, it is often difficult to collect business information from water vendors regarding their operations, especially data regarding financial particulars such as sales, costs, and profits, as vendors may guard such information as proprietary. Third, systematic sampling is rare, due to a lack of knowledge about the appropriate sampling frame for informal markets, and, as a result, the representativeness of findings is often in question. Most of the existing empirical work has been based on surveys of tanker truck operators and of distributing vendors (e.g., individuals selling water door-to-door in pushcarts) and rarely includes surveys of owners of private water vending businesses.

Kathmandu, Nepal was chosen for a case study because its booming water vending industry arose from two conditions common to many cities: a municipal utility that fails to deliver adequate water to its population over a significant period of time and a rapidly growing population that is partly dependent on alternative sources of water. As we document below, because private water sales comprise a significant proportion of the water supplied to end users in Kathmandu, the city provides an opportunity for in-depth research into the structure of a large urban water market. Many cities in the developing world exhibit similar features and issues. Lessons learned from the analysis presented here may assist policymakers in other cities in developing countries.

Drawing upon data collected in 2014 from in-person interviews with 120 water vendors and from respondents in 1,500 households, as well as from key informant interviews, this analysis presents an in-depth portrait of the structure of the informal water market in Kathmandu. We document the presence of five main types of water vending (commercial water source abstraction, tanker truck delivery, bottled water production, household delivery by distributing vendors, and sale of both bulk and bottled water by retail outlets such as neighborhood kiosks and private shops) and show that these functions may be consolidated in various ways. Financial analysis of the revenues and costs of commercial water source vendors and tanker truck vendors reveals that these businesses do not appear to be earning monopoly rents. That is, these components of the water vending supply chain appear reasonably competitive.

Background and study area

Although Nepal is one of the least urbanized countries in the world (only 17% of its population lives in urban areas), it is urbanizing rapidly. The urban population has increased 6% annually since the

1970s. This growth has occurred within a context of political instability. The first local elections in more than two decades were held in May 2017. During the lengthy preceding interval, Kathmandu, the nation's capital city had not had an elected local government but was run by officials of the central national government. That government was unable to cope with extreme unplanned urban development that had been exacerbated by an influx of migrants from rural areas, due in part to the decade-long Maoist insurgency (1996–2006). Kathmandu became a sprawling urban metropolis, with high levels of pollution, inadequate water and sanitation systems, increasing traffic congestion, and a growing energy crisis.

The Kathmandu Valley lies upstream of the Bagmati River Basin, which is regarded as one of the most water-stressed basins in the country (Pandey *et al.*, 2010a, 2010b). Groundwater has long been and continues to be an important source of water supply in the area. Significant groundwater abstraction started in the mid-1980s, when the Nepal Water Supply Corporation (NWSC) introduced groundwater into the municipal water supply system, and it became the most reliable source of water for the increasing urban population (Pandey *et al.*, 2009). Since 1986, the abstraction rate has exceeded the recharge rate, which has resulted in falling groundwater levels.

The municipal water supply system in Kathmandu is comprised of four distinct components:

1. the traditional water system, consisting of stone spouts, dug wells, tanks, and ponds that were built over several centuries;
2. the piped network system;
3. private water extraction by households and industries; and
4. various types of water vending.

The traditional stone spout system is a unique aspect of the water supply system in the valley. Historically, the spouts were connected to a network of canals that delivered water from upland sources to storage ponds. These ponds, in turn, recharged shallow groundwater aquifers that maintained a flow to the stone spouts or dug wells. A study conducted in 2007 revealed that of the 400 original stone spouts in the valley, only about half were still functional (UN-HABITAT, 2007).

The first piped water distribution network was introduced in 1891. At the time, its main purpose was to serve the families of the rulers and elites in Kathmandu. In 1928, the system was expanded to supply the general public. In 1972, the World Bank made a loan to the Nepalese Government to improve urban water supply and wastewater services. The Water Supply and Sewerage Board, formed in 1974, was renamed the Nepal Water Supply Corporation (NWSC) in 1989. In 2006, the Kathmandu Valley Water Supply Management Board Act was passed, which divided the NWSC into three separate organizations. Since then, the Kathmandu Upatyaka Khanepani Limited (KUKL) has been designated as the sole service operator and responsible for providing water and sewerage services to the people of the entire valley.

Between 2001 and 2011, the population of Kathmandu increased from 1.1 to 1.7 million (CBS, 2012), but very little corresponding investment has been made in the piped water system, which led to a substantial gap between the supply and demand. In 2014, water demand in the valley was estimated to be 360 million liters per day (MLD). The public water utility's current production reaches only about 76 MLD in the dry season and 123 MLD in the wet season (KUKL, 2014). In addition, annual losses from the municipal piped water supply system have been estimated to be over 70% (Dixit & Upadhya, 2005).

The result of this state of affairs is that as of 2014, less than 20% of the population was receiving a reliable supply of piped drinking water (ADB, 2010). To overcome the shortages, a long-term investment program is in progress to divert water from the Melamchi River outside the valley and deliver it to Kathmandu via a long (26 km) tunnel (Dixit & Upadhya, 2005). This project has experienced protracted delays. The current gap between demand and supply in Kathmandu is being met primarily through private groundwater extraction by households and vendors and through sales of water from a variety of private water vendors.

Research design and fieldwork

For our research in Kathmandu Valley, we used a mixed-methods research approach that incorporated data from structured household and vendor surveys as well as from in-depth key informant interviews. This combined strategy made it possible to triangulate our findings from multiple sources and extrapolate these findings to the city as a whole. A mixed-methods approach seemed especially appropriate because only limited and dispersed knowledge was available about the various unregulated water vendors. All data collection activities were conducted in Kathmandu in 2014.

Adopting protocols for data collection from a similar study (Whittington *et al.*, 1989), we conducted a survey of 120 water vendors engaged in various combinations of three activities along the water vending supply chain: (1) commercial abstraction of bulk water from natural sources or bore holes (i.e., not from the piped system), (2) tanker truck delivery of bulk water to end users, and (3) bottled water production. In our sample, 'vendors with commercial water sources' thus refer to vending businesses that owned water sources and sold bulk water in large or small quantities from supplies other than the piped water system. Our descriptor 'vendors with tanker trucks' refers to businesses that used tanker trucks to deliver bulk water to households and businesses. Our third category, 'vendors with bottled water facilities,' are operations that produce bottled water, that is, treated drinking water packaged in 20-L plastic 'jars' or 1-L bottles. For purposes of analysis, 'integrated vendors' involved in more than one of these operations were assigned compound names.

For the water vendor survey, lists of commercial water source vendors, tanker truck vendors, and bottled water vendors were collected from the central offices of the vendors' respective business associations: the Valley Drinking Water Source and Tanker Entrepreneurs Association (VDWTEA) and the Nepal Bottled Water Industries Association (NBWIA). The commercial water source list was verified by field visits and consultations with the owners of different boreholes and raw water sources, such as springs[1]. From in-depth interviews with KUKL and the heads of the business associations, we determined that the associations' membership lists were the best sampling frame for the water business population, representing the majority of vendors operating in the Kathmandu Valley. According to these two lists, there are approximately 67 commercial sources of water, 700 tanker trucks supplying the city with water through 210 vendors, and 200 bottled water vendors selling drinking water through 20-L jars and 1-L bottles. From the lists, we drew three random samples: (1) 40 from the list of commercial water source vendors; (2) 40 from the list of vendors with tanker trucks; and (3) 40 from the list of bottled water vendors. Figure 1 shows the location of the vendors we interviewed.

---

[1] This list only included commercial water source vendors who sold to independent tanker truck vendors and did not have their own tanker trucks. Vendors with commercial water sources who had their own tanker trucks were included on the tanker truck association list. Some tanker truck vendors had their own sources as well, but they were not categorized as 'commercial water source vendors' if they only supplied water to their own tanker trucks.

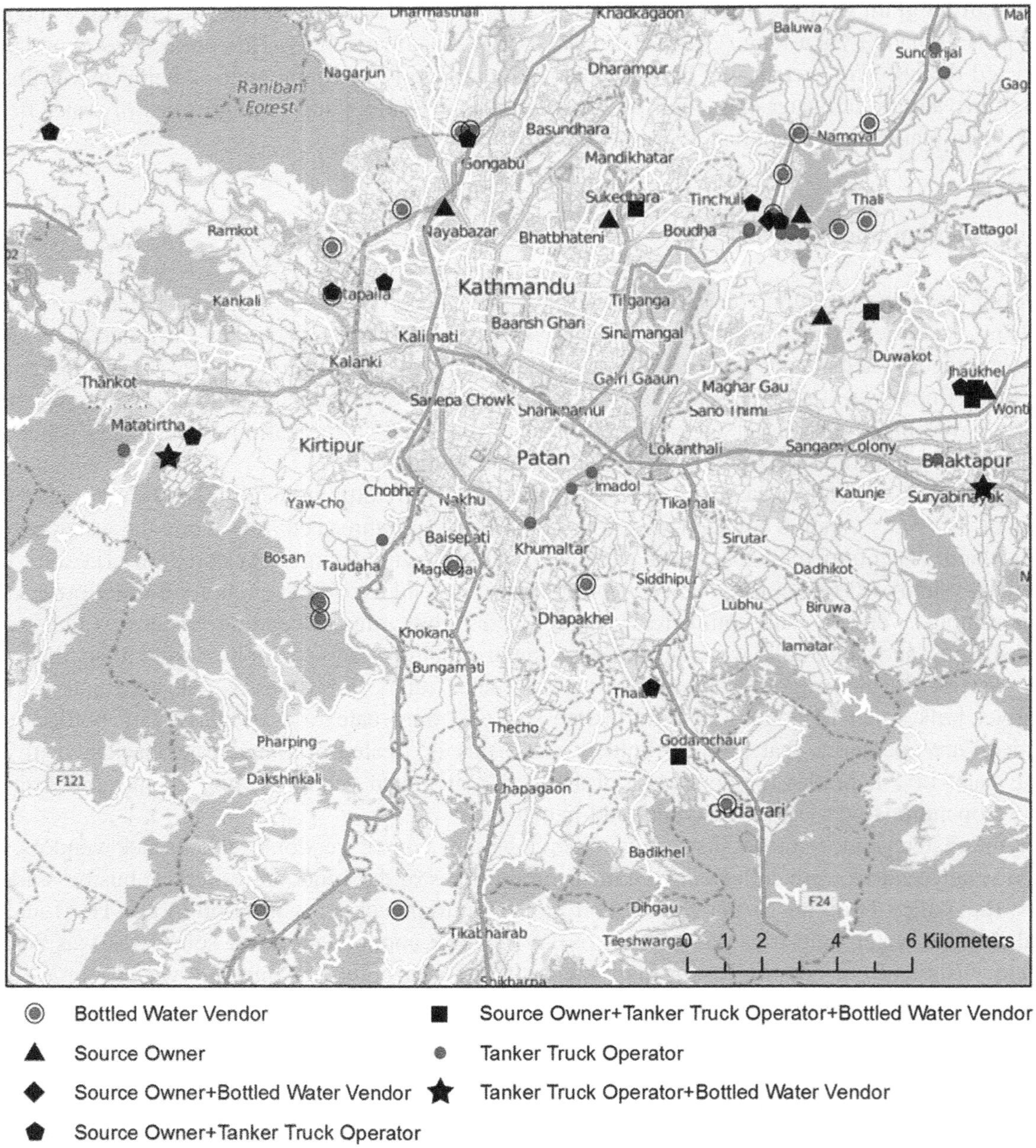

Fig. 1. Map of Kathmandu Valley, showing the location of vendors interviewed.

For all three types of vendors, we specifically wanted responses from the business owners. If an owner was unavailable, we instead solicited responses from the manager. Overall, of the 120 vendors we surveyed, 84% of the respondents were owners. Of the vendors with a commercial water source (along with any level of integration with other water business activities), 82% were owners. All of the vendors with only tanker trucks had owners as respondents. For vendors with only bottled water facilities, 70% of respondents were owners.

The water vending survey instrument presented questions that were common for all types of vendors, such as growth of their operations, customer numbers and types, relationship with the association (if a member), service details, operational and capital costs, and major challenges faced. The survey instrument also included three separate sections containing questions specific to the type of water vending (e.g., whether the vendor produced bulk water at a commercial water source, delivered bulk water with tanker trucks, and/or produced bottled water). The survey instrument was designed to allow for varying kinds and levels of consolidation among these three activities. We did not examine the distribution of bottled water or the sale of bulk water at retail outlets.

We also conducted a large household survey ($n = 1{,}500$), which included questions about the quantities of water a household purchased from different types of vendors and the prices paid (Gurung *et al.*, 2017). In addition, nine in-depth key informant interviews were conducted with government officials from KUKL and Kathmandu Valley Water Supply Management (KVWSM), the most recently elected mayor of Kathmandu, a retired university scholar, development practitioners, and the presidents of the two business associations (VDWTEA and NBWIA)[2].

The household survey covered respondents from 1,500 households across the five municipalities of Kathmandu Valley – Kathmandu itself, and Lalitpur, Madhyapur, Kirtipur, and Bhaktapur. These sample households had been previously surveyed in 2001 (Pattanayak *et al.*, 2005). In 2001, these household clusters were located using aerial maps provided by the Central Bureau of Statistics for the 1996/1997 World Bank Living Standard Measurement Survey for Kathmandu. In three of the five municipalities in the Kathmandu Valley (Kathmandu, Lalitpur, and Bhaktapur), a previously conducted complete enumeration of all households was used as the sample frame (SILT Consultants and Development Research and Training Center, 1999). For Kirtipur and Madhayapur, the 1991 population census was used as the sampling frame. Wards were then selected from the sampling frame on the basis of a probability-proportional-to-size sampling approach that ensured households had an equal opportunity of being included in the sample (Babbie, 1990). After a ward was selected for inclusion in the sample, sub-wards were drawn randomly. The final sample consisted of 60 clusters of 25 households each, covering all five municipalities in the Kathmandu Valley. In each cluster selected for inclusion in the sample, respondents from all 25 households were interviewed for the study. Because probability-proportional-to-size sampling depends on the size of population, some wards had more than one cluster in the final sample.

For the 2014 survey, if a household included in the 2001 survey could not be located, a nearby household in the same cluster was selected as a replacement. When a household head from the 2001 survey was unavailable, the present head or another responsible member of the household was chosen instead. In total, we were able to locate and re-interview a respondent in 927 of the 1,500 households visited in 2001. Thus, the 2014 survey included 573 replacement households.

The nine key informant interviews were audio-recorded with prior written permission obtained from the interviewees. In addition, written notes were taken during the interviews. Government officials and scholars were asked about the major milestones in the history of water infrastructure investment in Kathmandu Valley, major challenges faced in the development of urban water and sanitation services,

---

[2] Approval for the two surveys and for the informant questionnaire, and for their implementation protocols, was obtained from the National University of Singapore's Institutional Review Board.

institutional roles and responsibilities, and the growth of the private water market. We also discussed with key informants their interactions with the private water vendor market[3], access of vulnerable and poor populations to water and sanitation services, and impediments to expanding the piped network to all residents. Chairpersons of the two vendor associations were asked questions regarding the history and organizational structure of their respective organizations, their relationship and interaction with the government, and their main activities, including lobbying, as well as the factors that influenced whether water vendors joined their association.

The information obtained from the household and water vendor surveys was used to estimate the city-wide scale of quantities of water sold and of money paid and received at different points in the water vending supply chain. These estimates were cross-checked against information we had received from the informant interviews. In addition, we estimated the total water supplied by the public water distribution network, using water use data from our household survey and estimates of the total number of connections in the city ($\approx$195,000).

Detailed financial data were collected from commercial water source vendors and the tanker truck vendors. These data enabled us to construct monthly accounts of revenues and costs using standard financial accounting methods. By constructing a basic financial statement for each vendor, we were able to assess the company's profitability.

Results

Overview of water vending in the Kathmandu Valley

As the population of the Kathmandu Valley has increased, the water vending market has expanded rapidly. According to the president of the commercial source and tanking business association (VDWTEA), in the early 1990s, only 60–70 tanker trucks delivered bulk water in the valley, and no commercial water sources existed to supply tanker trucks or the bottled water vendors. The tanker truck vendors obtained water primarily from natural sources (springs and rivers). In 1994, the first commercial water source (a borehole) for use by tanker trucks was constructed in Jorpati, in Kathmandu Municipality, which was quickly followed by several additional similar enterprises. The rest of the bulk water sold by tanker truck vendors was still obtained from natural sources. The years 1996–2000 witnessed the peak of the Maoist insurgency and the great migration from rural areas to Kathmandu and over that short interval the water vending industry in Kathmandu boomed. The number of tanker trucks increased from 160 to 500, and many more commercial water sources (boreholes) were constructed, because many natural water sources had become depleted and contaminated. The president of the bottled water business association (NBWIA) reported much the same history.

By the time our fieldwork began in 2014, about 700 tanker trucks supplied the city with water obtained from 210 commercial water sources. Approximately 200 bottled water vendors were selling 20-L plastic jars and 1-L bottles to shops and families. A majority of the water vendors included in our vendor survey had started business after 2010. The oldest business included in the survey was established in 1992.

[3] For example, the public water utility had its own tanker trucks for distributing bulk water, thus competing with the private water tanker vendors. KVWSM decided key regulations, such as permits for commercial water sources. Ultimately, the associations were formed, organizing the various types of water vendors.

A great majority (88%) of the water vendors in our sample were members of one or the other of the two business associations. At time of the survey, the main role of the associations was to represent the interests and needs of the vendors to the government. There were no legal or institutional mechanisms that specify the price of vended water, quantities sold, or quality standards (Moench, 2001; Janakarajan & Moench, 2006). The business associations had issued pricing guidelines for their members, but these guidelines were not enforced. Private vendors were required to pay some taxes (such as the road and vehicle tax for tanker trucks, value-added tax (VAT) for bottled water producers, and income taxes). In addition, commercial water source vendors must obtain licenses to extract water from boreholes. Licensure abides by a national legal standard but often depends on the approval of the local community where a borehole is located (Shrestha *et al.*, 2012). There have been ad hoc instances of government testing of the quality of water sold by different types of vendors, but as yet no systematic public sector efforts have emerged to ensure that vended water meets recognized quality standards. As a result, individual vendors have created their own water quality standards to ensure that they retain the trust of their customers and to distinguish their product from those of competitors.

Until recently, and during the time of our survey, the national law requiring licensure for commercial water vendors was not enforced. Thus, apart from costs of extraction, commercial water source vendors did not pay a price for their raw water supply, either for surface or spring water or for groundwater. In 2018, growing concerns over groundwater depletion caused the national government to consider the introduction of a licensing framework for the use of commercial water sources. As the new framework was not yet enacted, the drilling of new boreholes was put on hold, resulting in a barrier to entry into the vending market. Aware of the vital role played by the vendors in providing water to underserved Kathmandu residents, authorities remained lenient with unlicensed water extraction from the existing commercial water sources.

How water and money flow through the water vending supply chain

Figure 2 illustrates the water vending system in Kathmandu and how water and money flow along some of the links in the supply chain[4]. Our calculations show that during the dry season of 2014, households and businesses purchased approximately 370,000 to 500,000 m^3/month[5] from the private water market, generating total revenues for commercial water source vendors, tanker truck vendors, and bottled water vendors of about US$1 million/month. This represents about 20% of the water used by households in Kathmandu in that dry season. Commercial water source vendors sold 269,000 m^3 of water each month and received US$103,000/month from households, businesses, and tanker truck vendors. Tanker truck vendors delivered and sold 371,000 m^3 of water and received US$806,000/month, of

---

[4] Whittington *et al.* (1991) presented the first such water-money flow diagram for an urban area (Onitsha, Nigeria), and Whittington *et al.* (1993) extended such analysis to sanitation-money flows in Kumasi, Ghana. The Sustainable Sanitation Alliance has promoted the use of similar flow diagrams (but without the money flows) to describe sanitation conditions in both urban and rural areas, and these 'shit flow diagrams' are now widely used by sanitation professionals in the Global South.

[5] Estimates of the quantities of vended water vary depending on whether one extrapolates using the vending sample or the household sample. If we use data from the vending surveys, the total estimate is about 370,000 m^3/month. If we use data from the household survey, the total is about 500,000 m^3/month.

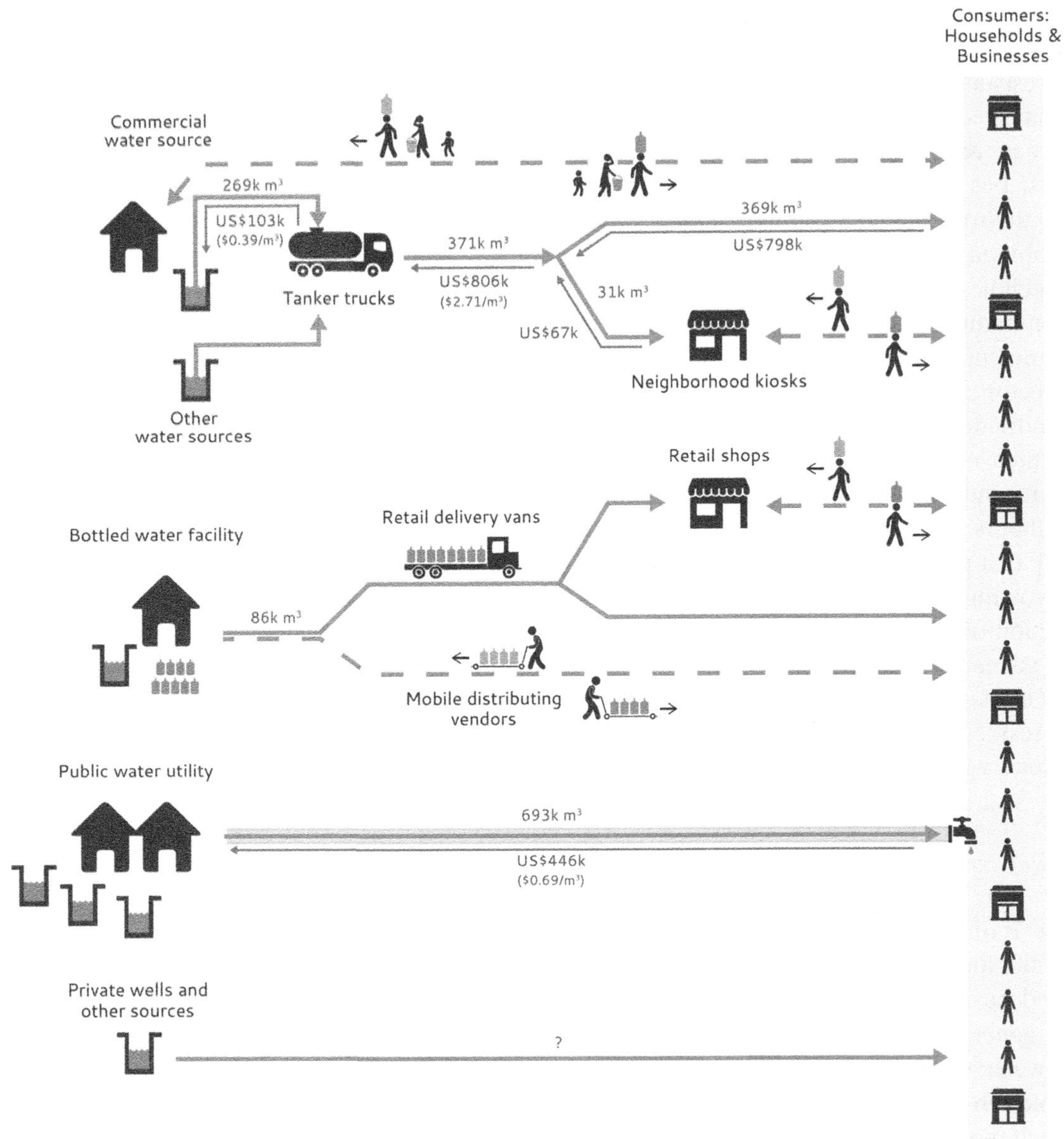

Fig. 2. Money and water flows in Kathmandu Valley, 2014 (dry season).

which 31,000 m^3 was sold to retail outlets[6]. The average price of bulk water sold by tanker truck vendors was US$2.17/m^3 at the time of our study.

---

[6] There is a discrepancy in our estimates of the amount sold by commercial water source vendors (268,701 m^3) and tanker truck vendors (370,860 m^3), as we would expect the amount sold by tanker truck vendors to be larger than that sold by commercial water source vendors. We speculate that this difference may be due to (1) tanker truck vendors drawing upon their own, noncommercial sources or (2) an underestimation of the number of commercial water source vendors.

From our household survey, we found that the public piped water network accounted for about a quarter of total household water use (693,000 m^3/month)[7]. For our sample of 1,500 households, private wells provided 45% of the total water respondents collected and used. The piped water network supplied 26% and vendors supplied 20%. The rest (about 10%) came from other sources, such as public wells, public taps, neighbors, rainwater, ancient stone taps, and surface water.

Structure of the water vending industry in Kathmandu Valley

Commercial water source vendors. Commercial water source vendors are at the beginning of the supply chain in the water vending industry (Figure 3). These vendors abstract water from natural sources or from boreholes. They sell water to tanker truck vendors, but most also give some water gratis to households. Most commercial water source vendors (87% of our sample) also treat this water before it is loaded into tanker trucks. The most common method of treatment is sand filtration.

Commercial water source vendors do not pay for their raw water supply but do incur capital costs for boreholes, pumps, and other facilities for delivering water to customers. Those who treat their raw water before selling to customers incur treatment costs as well. Other costs include electric, diesel, or petrol fuel for running pumps and other equipment, and labor costs for managing the facilities and sales.

In our sample, commercial water source vendors were clustered in nine areas in the Kathmandu Valley: Jorpati, Matatirtha, Swayambhu, Balaju/Bus-Park, Sundarijal, Chovar Chalnakhel, Jhaukhel, and Satdobato (Figure 1). Most (87%) offer free water to the communities within which their source is located. These vendors reported that they do so mostly to help their communities. However, our key informant interviews with government officials revealed that this policy was primarily pursued because over time, community resentment tended to grow regarding the private sale of a common resource. This rationale was corroborated in the interview with the president of the commercial vendors' business association (VDWTEA), who said that it had become extremely difficult to dig new boreholes as local people were no longer willing to give permission (as required by licensing authorities). Communities began to feel that the water source was no longer administered for the people but rather for the source vendor. Thus, the provision of free water to the local community had become the norm, and families came freely to the source with their own buckets and collected water for household use. Another common practice was for the commercial water source vendor to provide the police force with one free tanker load of water per week.

Commercial water source vendors have two main options for consolidating their operations with other components of the supply chain. First, they can purchase their own tanker trucks and use those trucks to expand into the delivery of bulk water directly to households and businesses. Or, second, they can take up bottled water vending, using their commercial sources as the raw water input into more sophisticated water treatment facilities. This treated 'drinking water' is then bottled into the common 20-L plastic jars and 1-L bottles for distribution to households and businesses. These two consolidation options are not mutually exclusive. Some vendors have consolidated all of these operations, covering the entire process from commercial water source extraction, to tanker truck distribution, to treatment, and bottling of drinking water.

[7] Calculated using the number of private water connections in Kathmandu Valley (194,718), as reported by KUKL. Our survey included 1,051 households using private water connections, and we calculated the average estimated volume of water collected from each connection based on household estimates.

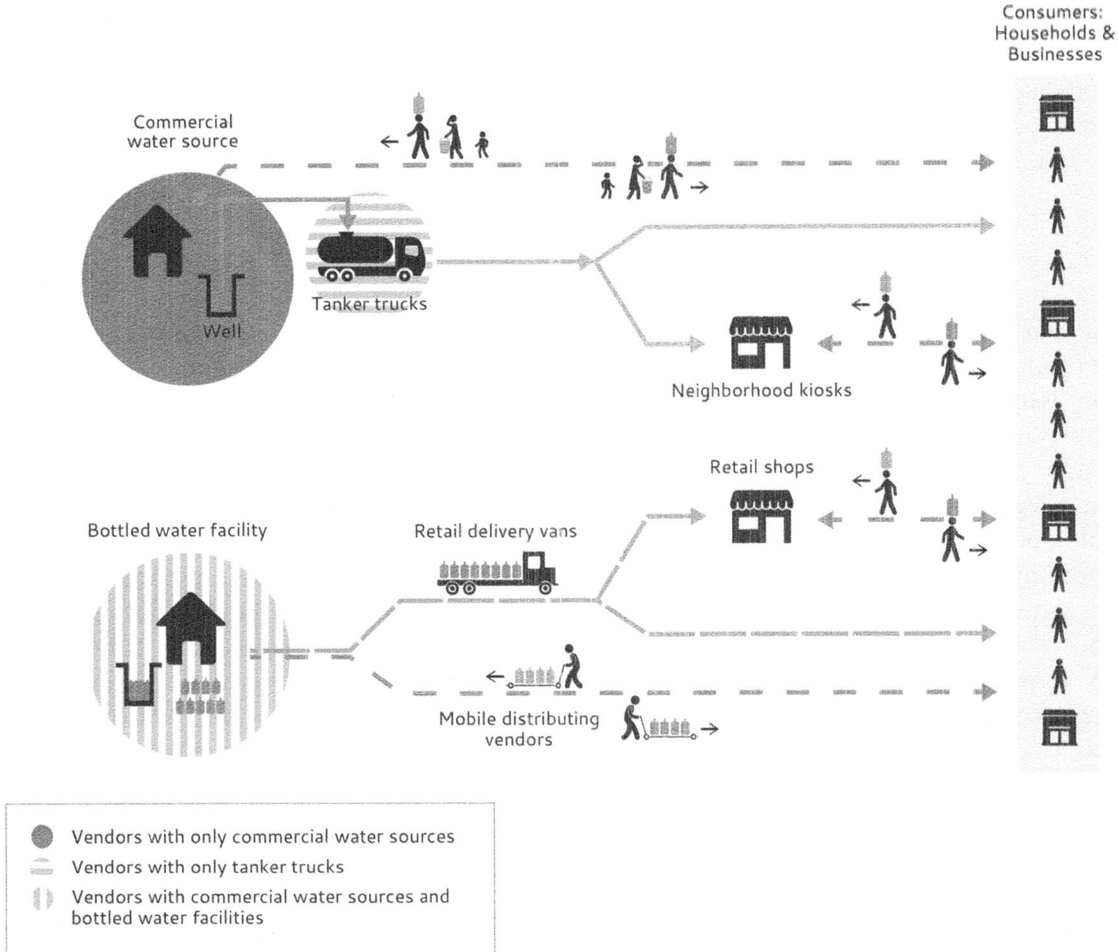

Fig. 3. Water vending industry structure.

Almost all the water from commercial water source vendors received by households in our study was delivered through tanker trucks. Only 15 of the 39 commercial water source vendors (38%) ran a source water-only operation, selling water to other, privately owned tanker trucks and to households. Another 46% of the commercial water source vendors in our sample had their own tanker truck operations. A few (3%) had their own bottled water operations, and 13% had both their own tanker trucks and bottled water facilities.

Tanker truck vendors. Vendors with tanker trucks collect water from commercial water source vendors and deliver it to both households and businesses (Figure 3). Households use bulk water for a variety of purposes. It may be used for drinking, but usually only after additional treatment (either boiling or filtering) by the households themselves. Two-thirds of the bulk water sales for tanker truck vendors in our sample were to households. Another 17% of sales were to businesses. Sales to neighborhood kiosks and to construction sites accounted for 7% each, and 2% went to hospitals.

The tanker trucks in businesses included in the survey were of varied capacities (5–15 m^3, predominantly 7 m^3). During the rainy season, each vendor sold about three to four tank loads per day, thus making three to four trips back to the commercial water source to refill. During the dry season, each sold approximately twice that amount, requiring seven to eight trips per day for refilling. During the rainy season, most tanker truck vendors were able to fulfill demand within about 4 h. The same process took much longer in the dry season, with customers waiting an average of 1–2 days for their order. A large majority of tanker truck vendors (79%) reported that in the dry season, they had more orders than they could fulfill.

The main operating costs for a tanker truck were fuel for the truck, and labor costs for the driver and an assistant, who helps with filling the tank at the source and discharging the tank at points of sale. A majority of tanker businesses (62%) had only one truck and employed only one driver and one driver's assistant. Owners of two trucks comprised most of the rest of the total (31%), and only a few (7%) had three trucks. A vendor who only deploys tanker trucks also incurs costs of purchasing water from a commercial water source vendor. Tanker truck vendors are also subject to taxes, fees, and regular maintenance and repairs on their trucks. Their only capital costs are expenditures to purchase their tanker trucks and for major repairs.

A tanker truck vendor has two main options for further integration with the water vending supply chain. The first is to integrate backward along the supply chain and drill (or purchase) a commercial water source, thus both selling bulk water and self-supplying tanker(s). In our sample, 30% of the vendors with tanker trucks had their own source of water (not necessarily a commercial water source), about half from groundwater and half from surface water sources. Second, the vendor can integrate forward in the chain and begin selling bottled water (Figure 4(b)). This combined strategy enables the tanker vendor to supply bulk water to households and other businesses via tanker trucks and also to deliver higher-quality bottled drinking water directly to households and businesses (using different delivery vehicles).

Bottled water vendors. Vendors that produce high-quality bottled drinking water necessarily have a proprietary source for their water. Some may use natural water sources (e.g., springs), and others may use groundwater (their own boreholes)[8]. Bottled water vendors often sell 20-L plastic jars of drinking water to mobile distributing vendors who, in turn, deliver them to households and businesses.

The main capital costs for a bottled water vendor are the water treatment and bottling facilities, and the land and building needed for these operations. They primarily use reverse osmosis technology to treat their raw water. Many in our survey used automated bottling and packaging machines (see photographs in Appendix 4, available online). Bottled water facilities also have labor costs for running and maintaining their equipment and handling sales of their product.

A bottled water vendor can expand by establishing its own retail distribution network, delivering water directly to households, often with motorized delivery vehicles and with its own employees. Alternatively, it can expand by supplying bulk water and operating its own fleet of tanker trucks, and/or acquire and sell bulk water from its source.

Mobile distributing vendors. Our study did not include interviews with mobile distributing vendors that sell bottled water. We did learn from our household respondents and bottled water vendors that mobile distributing vendors purchase 20-L jars of water from bottled water facilities and transport these to households. In 2014, the price of a 20-L jar water was most commonly US$3.15. This price included a refundable deposit for the jar upon return. These 20-L jars can be refilled, at a price ranging

[8] It is possible that some use the public piped water distribution system.

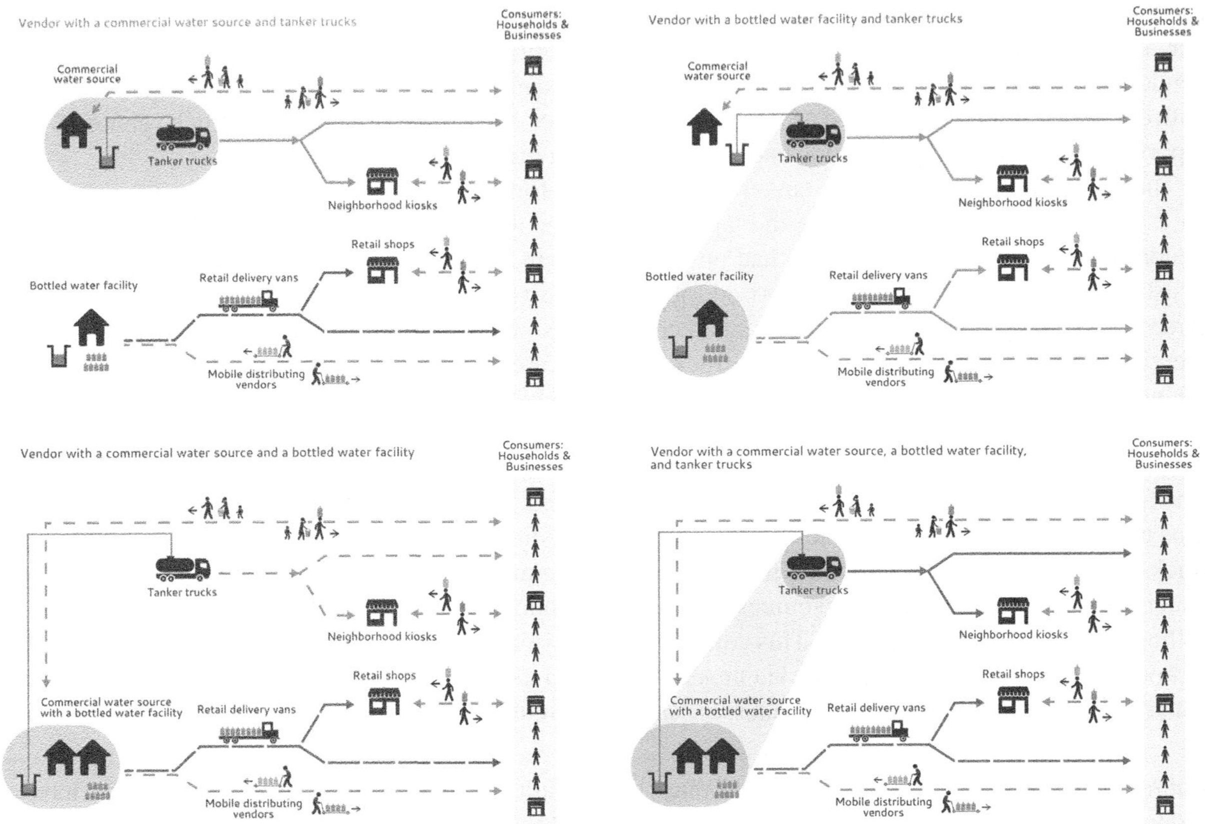

Fig. 4. Water vending consolidation options.

from US$0.10 to US$0.74, the most common amount being US$0.32. If the household owned its jar, the mobile distributing vendor delivered a full jar and picked up an empty one.

Retail outlets: kiosks and stores. Both bulk water and bottled water are also sold to households through retail outlets. Tanker trucks may deliver water to neighborhood kiosks, which then sell water to households in smaller quantities. The tanker truck vendor fills the kiosk's large storage containers, and from these, the kiosk vendor fills smaller containers for customers. Bottled water vendors likewise sell water to retail outlets, usually to small neighborhood stores, where households can walk to purchase 20-L jars and 1-L bottles and carry them home, or can request home delivery.

Entry into the market

The oldest vending business in the sample (a tanker truck vendor) began operations in 1992. Others began to enter the market in the early 2000s. The number of vendors entering the business was greatest in 2009 and 2010, then declined up until 2014, the year of the survey. Half of the source water and bottled water vending enterprises were owned by one person. Most (79%) of the tanker truck vending businesses in our sample also were owned by one person.

Average start-up costs for a commercial water source vendor were US$76,000 (median of US$49,000 and adjusted for inflation). For a tanker truck vendor, average start-up costs were US$27,500 (median of US$20,600). For a vendor with both a commercial source and tanker trucks, average start-up costs were US$87,000 (median US$69,000). Most vendors (74%) needed to borrow money to start their businesses. Most such start-up loans came from banks (59%) or cooperatives (18%), both of which charged a median interest rate of 5% for both banks and cooperatives (range 2–10%). Friends and families also provided loans, commonly with 5% interest as well.

Markets along the supply chain

The two main products delivered through the water vending supply chain – bulk water from natural sources or boreholes, and drinking water produced by bottled water facilities – were sold in four main markets.

The first of these (market 1) was the sale of bulk water from commercial water source vendors to tanker truck vendors (vendor–vendor) (Figure 3). In 2014, commercial water source vendors charged tanker trucks US$2.15 on average to fill for a 5,000 L tanker truck and US$4.86 for a 14,000 L truck. Averaging over different size tanker trucks, commercial water vendors charged tanker truck vendors an average of US$0.38/m^3 (Table 1).

Market 2 was the sale of bulk water from tanker trucks to households and businesses (vendor–end user sale). Tanker truck vendors charged households and businesses an overall average of US$2.25/m^3, about six times the price they paid the commercial water source vendors. Charges were, on average, US$13.78 for a 5,000 L tank to US$24.73 for a 14,000 L tank. Although demand for tanker truck water was much higher in the dry season than in the rainy season, tanker truck vendors charged the same price in both seasons. It is not clear to us why they did not to raise their prices in the dry season.

Table 1. Description of four markets for vended water (prices in the dry season).

Product	Bulk water		Bottled water	
Market	Upstream market	Consumer market	Upstream market	Consumer market
Sellers	Source owners	Tanker truck operators[a]	Bottled water operators	Bottled water operators, distributing vendors
Buyers	Tanker truck operators	Businesses, households	Distributing vendors	Businesses, households
Price (US$/m^3)				
Mean	0.38	2.25		
Standard deviation	0.08	0.57		
10th percentile	0.29	1.75		
Median	0.38	2.19		
90th percentile	0.50	2.74		

[a]Our numbers are calculated using tanker truck sales to consumers. Very few source owners sell directly to households. Large businesses often will have their own water tanker trucks, which would purchase water directly from the source.

Market 3 was the sale of drinking water produced by bottled water facilities to mobile distributing vendors for subsequent purchase by households and businesses (vendor–vendor). Our research protocol did not collect information permitting an estimate for the price at which bottled water vendors sold water to mobile distributing vendors. From interviews, we learned that some bottled water vendors delivered to distributing vendors for a delivery charge of between US$13 and US$16 for a minimum order of 150 20-L jars.

Market 4 was the sale of 20-L jars delivered by mobile distributing vendors and bottled water vendors directly to households and businesses (distributing vendor–end user). In 2014, the price for household delivery and exchange of a 20-L refill jar were US$0.47–US$0.63, depending on water quality and company policy. Additional market transactions occurred at kiosks for bulk tanker truck water and at neighborhood retail outlets for 20-L jars.

From the households' perspective, the water vending supply chain offered two products and six main purchase options. (1) A household member could walk to a commercial water source vendor and collect bulk water for free. (2) The household could pay a tanker truck to deliver the same quality bulk water directly to the home, again purchasing in large quantities. (3) In some neighborhoods, households could purchase this same quality of water from kiosks in smaller quantities. (4) A household (or business) could travel to a bottled water vendor and collect 20-L jars of drinking water (with a minimum purchase). (5) A household could pay either the bottled water vendor or an independent mobile distributing vendor to deliver 20-L jars of water to the home. (6) A household could purchase 20-L jar(s) or 1-L bottles from neighborhood retail outlets and carry the water home themselves.

Figure 4 summarizes the four main ways that water vending entrepreneurs could integrate activities to serve more than one of these markets:

1. Commercial water source vendors could purchase their own tanker trucks in order to deliver water straight from the source directly to households, selling in both Markets 1 and 2. We found that this was a common integration strategy.
2. Commercial water source vendors could expand their product line by integrating bottled water vending, thus selling in Markets 1, 3, and 4.
3. A tanker truck vendor could integrate operations with a bottled water vendor, thus selling in Markets 2, 3, and 4. In 2014, this option was uncommon in the Kathmandu Valley. Only two of the vendors in our sample had adopted this integration strategy.
4. A bottled water vendor could integrate a retail distribution of 20-L bottled water into its operations, thus eliminating sales in Market 3, selling its product only in Market 4.

Additional options might have included even more aggressive integration strategies, but to the best of our knowledge, these have not yet materialized in Kathmandu.

Most of the commercial water source vendors (100%), tanker truck vendors (78%), and bottled water vendors (56%) in our sample stated that they adhered to the prices set by their business associations. However, when questioned further, the tanker truck vendors admitted they often reduced prices to fight off competition. When tanker truck vendors were asked how they priced the water they sold, 82% stated that their business association rules did not factor into their pricing decisions. In contrast, it seems that bottled water vendors used water quality instead of price to differentiate their product from that of competitors.

Financial accounts for commercial water source vendors and tanker truck vendors

Our data allowed us to take a more detailed look at the revenues, costs, and profits by constructing income statements for three types of vendors in the supply chain (Markets 1–3). We could then assess performance of the three types of vendors involved: (1) commercial water source vendors selling water to tanker trucks (Market 1); (2) tanker truck vendors without other vending activities (Market 2); and (3) vendors that owned both a water source and tanker trucks (Markets 1 and 2). (We did not create income statements for participants in Market 4, bottlers and mobile distributors.) To construct the income statements for sellers in Markets 1, 2, and 3, we calculated the monthly revenues, financial costs, asset productivity, and three different measures of profitability for each enterprise, in both the dry and wet seasons (Tables 3, 4 and 5). Appendices 1, 2, and 3 (available online) present the full set of vendor-specific results.

From these income statements, we could assess the profitability of each business and compare businesses with each other. Profitability is arguably the most important criterion for evaluating the performance of a firm (Smith *et al.*, 1998). Profitability metrics, which measure the return that the firm's owners receive from their investments, have been widely used in research on information systems, strategic management, and finance (Smith *et al.*, 1998). There are four key metrics: return on assets (EBITDA/average assets), gross margin (direct product costs/revenue), operating margins (EBITDA/revenue and EBIT/revenue), and net profit margin (net income/revenue) (Smith *et al.*, 1998; Agrawal & Hall, 2014). Return on assets (ROA) is a commonly used metric for overall company performance (see Hunton *et al.*, 2003; Andres, 2008; Adams *et al.*, 2009). Our calculations for each vendor's income statement began with an estimate of total monthly revenue. From this, we estimated four different measures of profit: (1) gross margin; (2) earned income before interest, taxes, depreciation, and amortization (EBITDA); (3) earned income before interest and taxes (EBIT); and (4) net income.

The first, gross margin, shows the profit made from making and selling a product, without taking into account overhead and other fixed costs. Gross margin is a reflection of how much value to the basic inputs is added by the company utilization of capital and technology. In competitive markets, gross margins should be similar across firms, because companies with larger gross margins can lower their prices to outcompete companies with smaller gross margins. Gross margin is calculated by subtracting the cost of goods sold from monthly revenue. The cost of goods sold include only direct costs attributable to the product, including material input costs (or product costs), transportation costs (or supply chain costs), and direct labor costs. These are all variable costs; in other words, they are a function of the quantity of the product sold. For example, for tanker truck vendors, the cost of purchasing bulk water from a commercial source counts as a product cost; the fuel cost of transporting water from the source to the consumer is one component of the supply chain costs; salaries of truck drivers and truck driver assistants are direct labor costs. Our survey did not allow for direct attribution of cost of goods sold for source vendors and bottled water vendors because we did not ask these vendors how much was spent per unit produced (e.g., variable pumping costs and costs of each bottle for bottling). While we inquired about the cost of diesel, petrol, and electricity, we were unable to separate out the proportion of those costs used to pump water from the source (what would be attributed to product costs). Additionally, we did not inquire about direct labor costs for the source and bottled water vendors.

The second measure of profit, earned income before interest, taxes, depreciation, and amortization (EBITDA), provides a measure of the amount of money a company makes before accounting for the costs of capital assets and debt. It provides an idea of a company's profits independent of decisions about location (which determines taxes) and investment (in capital equipment), and it disregards sunk costs (i.e., costs already incurred that are difficult to reverse, such as drilling a borehole). EBITDA is calculated by subtracting from the gross margin the costs of selling the product, and general and administrative costs. These costs include overhead and management, maintenance and repairs, rent and utilities, and marketing and other sales costs. In our calculations, for the vendors with tanker trucks, we included, in the overhead and management category, any labor costs of staff not associated with the direct operation of tanker trucks. For source vendors, we were unable to separate direct labor from indirect labor costs and deducted all labor costs from the gross margin. For maintenance and repairs, we used estimates of monthly spending on equipment breakdowns, replacement of parts, and maintenance of trucks, equipment, pumps, and other equipment. For rent and utilities, we included the costs of renting the business premises and any other related buildings, and utility costs (electricity). We did not impute any costs for land and buildings that a business already owned. The costs of any permits, association membership, and insurance were included in 'marketing and other sales costs.'

The third measure, earned income before interest and taxes (EBIT), deducts from gross margin the costs of the company's capital assets. This is the operating income – that is, profits from operations, independent of debt or taxes. Debt structure can change depending upon how owners decide to raise money, access to loans, etc., and has little to do with the day-to-day operations of the business. EBIT is calculated by subtracting depreciation and amortization from EBITDA. To calculate depreciation costs, we used straight-line depreciation based on typical useful lives of different assets (see Table 2). We did not amortize intangibles such as software, etc. (Stickney *et al.*, 2010).

The fourth measure, 'net income,' shows the amount of money a company makes after deducting from revenues all costs (variable, fixed (or overhead), and sunk (difficult to reverse)), as well as taxes and interest on debt. This estimate is obtained by subtracting interest and taxes from EBIT. The interest included in our calculations was the interest on company start-up loans that were still being paid off during the time of the survey. Our survey did not ask about loans taken out after the inception of a company, so our reported interest paid constitutes a lower bound. Taxes included the VAT, income tax, vehicle and road taxes, and a village development committee (VDC) tax.

Table 3 shows that water source vendors without their own tanker trucks had small mean (US$699) and median (US$614) monthly gross margins in the dry season and reduced mean (US$310) and median (US$359) monthly gross margins in the rainy season. Estimates of mean and median monthly net

Table 2. Useful life estimates of capital assets.

Asset	Useful life
Buildings	20
Water storage tank/reservoir	15
Bottling and packaging	15
Equipment for bottled water treatment	15
Computers	5
Filters	5
Generator	15
Pumps and piping	10
Delivery van	10
Others: CCTV camera	5
Tanker truck	20

Table 3. Profits of vendors with only commercial water sources (dry and rainy seasons, US$/month).

	Mean	SD	10th percentile	Median	90th percentile
Dry season					
Revenue	942	673	404	771	2,053
Cost of goods sold					
Product costs	16	19			
Supply chain costs	228	278			
Labor costs (direct)	0	0			
Gross margin	699	692	−173	614	1,723
Selling, general, and administrative costs					
Overhead and management	95	113			
Maintenance and repairs	47	58			
Occupancy and real estate	151	106			
Marketing and other selling costs	7	10			
EBITDA	398	603	−307	296	1,254
Depreciation	415	483			
EBIT	−17	864	−1,198	75	1,058
Interest	104	252			
Taxes	6	9			
Net income	−27	852	−1,213	−195	1,048
Wet season					
Revenue	554	336	177	556	1,067
Cost of goods sold					
Product costs	16	19			
Supply chain costs	228	278			
Labor costs (direct)	0	0			
Gross margin	310	430	−271	359	883
Selling, general, and administrative costs					
Overhead and management	95	113			
Maintenance and repairs	47	58			
Occupancy and real estate	151	106			
Marketing and other selling costs	7	10			
EBITDA	9	411	−473	−10	612
Depreciation	415	483			
EBIT	405	574	−1,169	−393	371
Interest	104	252			
Taxes	6	9			
Net income	−516	521	−1,184	−574	290

income proved negative for both the dry (−US$127, −US$195) and rainy (−US$516, −US$574) seasons. Figure 5(a) presents a distribution of monthly net incomes of vendors with only commercial water sources. More than half were losing money. These results suggest that commercial water source vendors without their own tanker trucks are not earning monopoly rents and appear to be operating in a very competitive market.

Table 4 presents estimates of the same measures for tanker truck vendors that were not integrated with a commercial water source. Tanker truck vendors had somewhat higher mean and median monthly gross margins in both the dry season (US$846, US$749) and the rainy season (US$176, US$228). However, our estimates of mean and median monthly net income approached zero in the dry season (US$148,

Table 4. Profits of vendors with only tanker trucks (dry and rainy seasons, US$/month).

	Mean	SD	10th percentile	Median	90th percentile
Dry season					
Revenue	3,836	2,230	1,971	3,057	6,859
Cost of goods sold					
Product costs	693	459			
Supply chain costs	1,656	1,464			
Labor costs (direct)	641	575			
Gross margin	846	1,268	−325	749	2,426
Selling, general, and administrative costs					
Overhead and management	99	90			
Maintenance and repairs	266	115			
Occupancy and real estate	9	36			
Marketing and other selling costs	45	23			
EBITDA	427	1,306	−683	338	2,131
Depreciation	189	156			
EBIT	238	1,263	−991	193	1,762
Interest	58	86			
Taxes	32	37			
Net income	148	1,256	−1,073	83	1,541
Wet season					
Revenue	2,035	1,175	1,061	1,629	4,093
Cost of goods sold					
Product costs	373	244			
Supply chain costs	1,051	969			
Labor costs (direct)	434	345			
Gross margin	176	809	−814	228	769
Selling, general, and administrative costs					
Overhead and management	91	91			
Maintenance and repairs	266	115			
Occupancy and real estate	9	36			
Marketing and other selling costs	45	23			
EBITDA	−235	837	−1,352	−90	493
Depreciation	189	156			
EBIT	−424	844	−1,552	−241	375
Interest	58	86			
Taxes	32	37			
Net income	−515	845	−1,564	−373	301

US$83) and were negative in the rainy season (−US$515, −US$373). Table 4 also presents 90th percentile statistics for each variable. The measures for 90th percentile monthly gross margin (US$2,426 in the dry season) and 90th percentile monthly net income (US$1,541 in the dry season) are positive, suggesting that some tanker truck vendors were profitable. Vendors with more trucks had higher mean and median net incomes. Figure 5(b) shows the distribution of monthly net incomes for tanker truck vendors, which is centered slightly above zero. Together, these results suggest that tanker truck vendors as a whole were not earning monopoly rents and, like the commercial water source vendors, appear to have been operating in a competitive market.

Table 5. Profits of vendors with commercial water sources and tanker trucks (dry and rainy seasons, US$/month).

	Mean	SD	10th percentile	Median	90th percentile
Dry season					
Revenue	5,504	2,640	2,451	5,507	8,046
Cost of goods sold					
Product costs	144	231			
Supply chain costs	1,410	872			
Labor costs (direct)	686	283			
Gross margin	3,264	2,465	823	2,984	5,418
Selling, general, and administrative costs					
Overhead and management	216	247			
Maintenance and repairs	499	1,000			
Occupancy and real estate	301	294			
Marketing and other selling costs	49	38			
EBITDA	2,200	2,619	−1,637	2,116	4,895
Depreciation	604	465			
EBIT	1,596	2,457	−1,885	1,412	4,331
Interest	126	168			
Taxes	39	34			
Net income	1,430	2,486	−2,187	1,271	4,300
Wet season					
Revenue	3,093	1,704	1,364	2,665	5,457
Cost of goods sold					
Product costs	97	139			
Supply chain costs	991	497			
Labor costs (direct)	484	207			
Gross margin	1,521	1,579	−39	773	4,108
Selling, general, and administrative costs					
Overhead and management	212	248			
Maintenance and repairs	499	1,000			
Occupancy and real estate	301	294			
Marketing and other selling costs	49	38			
EBITDA	460	2,115	−1,787	298	3,308
Depreciation	604	465			
EBIT	−145	2,004	−2,035	−147	2,862
Interest	126	168			
Taxes	39	34			
Net income	−310	2,012	−2,336	−373	2,783

Table 5 presents estimates of the same financial measures for commercial water source vendors with their own tanker trucks. The average monthly dry season revenues of these vendors (US$5,504) are higher than monthly revenues of tanker truck vendors (US$3,836) or commercial water source vendors without tanker trucks (US$942). These vendors also had higher mean and median monthly gross margins in both the dry season (US$3,264, US$2,984) and the rainy season (US$1,521, US$773). However, although estimates of mean and median monthly net incomes are positive in the dry season (US$1,430, US$1,271), they become negative in the rainy season (−US$310, −US$373 per month). Table 5 also presents 90th percentile statistics. The 90th percentile measures for both monthly gross margin (US$5,4180) and monthly net income (US$4,300) in the dry season are positive and substantial. As

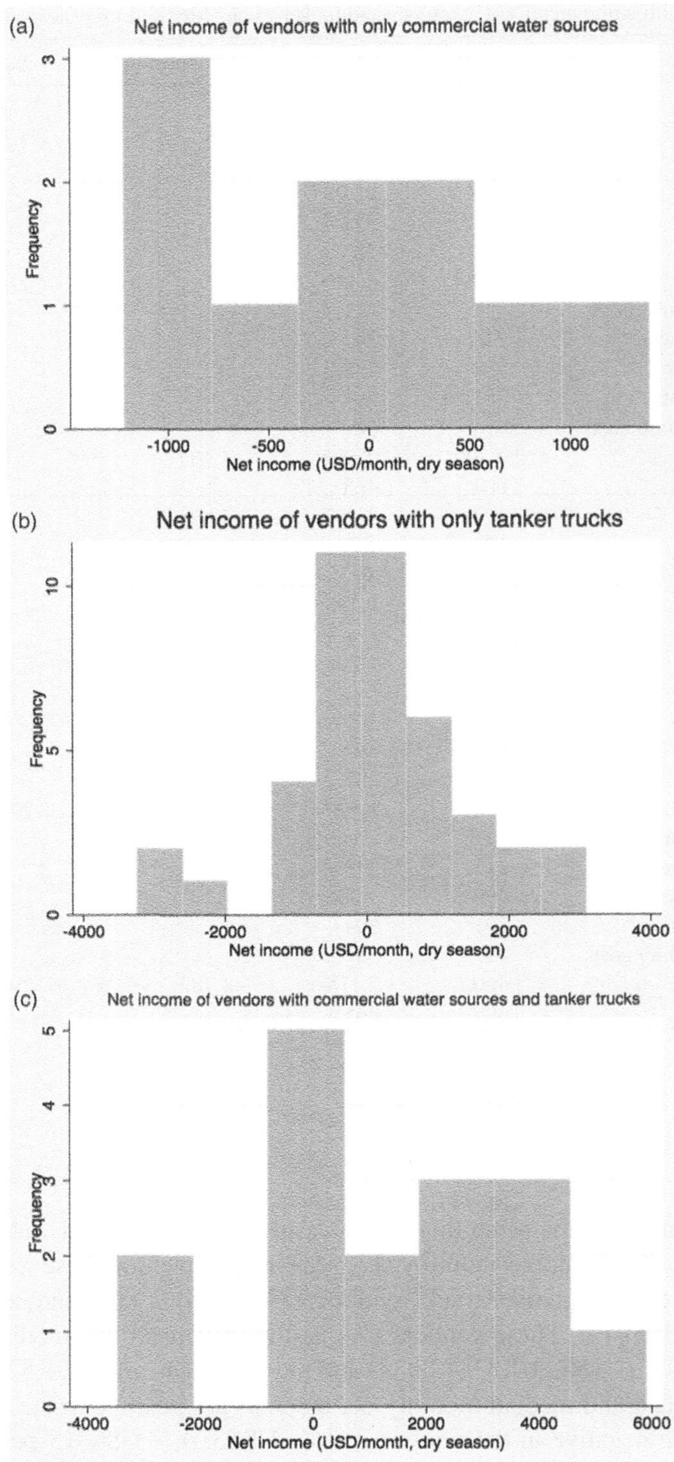

Fig. 5. Distribution of vendor net income by type.

Table 6. Firm productivity and profitability.

	ROA (%)	Gross margin/revenue		EBITDA/revenue		EBIT/revenue	
		Dry	Rainy	Dry	Rainy	Dry	Rainy
Commercial source vendor							
Mean	0.93	0.64	0.37	0.26	−0.37	−0.49	−1.23
SD	2.57	0.55	0.74	0.55	1.03	1.31	1.27
10th percentile	−0.82	−0.33	−0.83	−0.61	−1.77	−2.35	−2.34
Median	0.23	0.85	0.67	0.46	0.06	0.05	−2.00
90th percentile	4.70	0.96	0.95	0.76	0.67	0.73	0.55
Tanker truck vendor							
Mean	0.71	0.23	0.10	0.10	−0.14	0.05	−0.24
SD	10.87	0.33	0.44	0.35	0.48	0.35	0.49
10th percentile	−3.22	−0.05	−0.25	−0.21	−0.41	−0.31	−0.56
Median	0.96	0.25	0.14	0.10	−0.07	0.06	−0.17
90th percentile	8.47	0.59	0.54	0.42	0.26	0.39	0.17
Vendor with commercial source and tanker truck(s)							
Mean	3.43	0.53	0.38	0.32	−0.06	0.21	−0.28
SD	6.83	0.22	0.28	0.38	0.82	0.39	0.86
10th percentile	−5.78	0.31	−0.03	−0.45	−0.71	−0.52	−1.03
Median	2.07	0.51	0.41	0.36	0.12	0.25	−0.08
90th percentile	13.37	0.80	0.75	0.71	0.66	0.63	0.55

with stand-alone tanker truck vendors, these results suggest that some commercial water source vendors with their own tanker trucks were profitable. These findings further suggest that integrating commercial water source vending with tanker truck vending does result in increased market power. Figure 5(c) shows that most of the vendors with commercial water sources and tanker trucks were earning net profits, with only two vendors losing significant amounts of money. Of the three types of firms, the integrated commercial water source and tanker truck vendors appear from our estimates to have been the most profitable as a whole.

Table 6 summarizes vendor productivity and profitability. The average return on assets (EBITDA/assets) was less than 1% for both commercial source vendors and tanker truck vendors. However, results for vendors with both commercial sources and tanker trucks showed greater mean (3.4%) and median (2.1%) return on assets. Results for gross margins show that tanker truck vendors (Market 2) had the highest production costs, due to labor (drivers and assistants) and fuel costs, as their gross margins were the lowest (a mean of 0.23 in the dry season). Vendors with both commercial sources and tanker trucks had larger gross margins that were also more consistently positive. The standard deviation is low (0.22 compared to a mean of 0.53), indicating that the integrated vendors appear to have been operating in a competitive market. Operating margin without depreciation (EBITDA/revenue) illustrates the firm's profitability at its current level of assets. Integrated vendors with both commercial sources and tanker trucks are shown once again to be the most profitable of the three market types (0.32, compared to 0.10 for tanker truck vendors and 0.26 for commercial source vendors). Operating margin with depreciation takes into account the firm's capital investment decisions (such as the purchase of additional trucks and pumps) and reveals that many commercial source vendors were losing money, and tanker truck vendors were not profitable (with a mean operating margin of 0.05), whereas integrated vendors with commercial sources and tanker trucks remained profitable with a mean operating margin of 0.21. We did not include net profit margins in these calculations because of incomplete collection of data for interest on loans.

In summary, while there are some regulatory barriers to entry into Markets 1 and 2, such as restrictions on issuance of licenses for source vendors, these two markets appear to be quite competitive. Our estimates of net income are not excessive. There are no constraints on commercial water source vendors or tanker truck vendors regarding how much water can be extracted and sold, where their products can be sold, to whom their products can be sold, or the prices at which water can be sold. There seem to be many independent tanker truck vendors present in the market, and households purchasing from tanker trucks report that they are not dependent on any one vendor.

Concluding remarks

Our research shows that at the time of our study (2014), water vendors in the Kathmandu Valley operated a diverse, heterogeneous group of businesses. There was a supply chain with two main products: bulk water and bottled drinking water. Transactions occurred in four main markets: two upstream markets (between water source owners and tanker trucks, and between bottled water vendors and distributing vendors) and two consumer markets (between tanker truck vendors and consumers and between distributing vendors and consumers).

Each type of water vending business faced its own unique operational challenges and competitive pressures. Revenues, costs, and profits varied along the supply chain depending on the type, size, and integration of business operations. The characteristics of buyers and sellers also varied. Some water vendors were both buyers and sellers of water. Some water vendors were vertically integrated in the sense that they were involved in different phases in the supply chain, while others focused on only a single activity.

The portions of the water vending supply chain that we examined were all quite competitive. Yet, the fact that tanker truck prices for water did not respond to substantial changes in demand between the dry and wet seasons remains puzzling. Our research also discovered two active professional associations representing different types of water vending businesses in Kathmandu. But the influence of these associations on the behavior of different types of water vendors seems to have been modest; many water vendors reported that they did not follow the pricing guidelines promoted by the associations. It appears that competitive pressures are too strong for the associations to exert price control.

Understanding the structure and complexity of the water vending supply chain is a necessary first step in developing appropriate policy responses to improve the performance of informal water markets in any area and economy where they play a part. Policy interventions such as designing governance structures and regulatory frameworks may be needed in the future to address potential negative welfare consequences resulting from water vending (Collignon & Vézina, 2000; Conan & Paniagua, 2003; Kjellén & McGranahan, 2006; Banerjee, *et al.*, 2011). Such interventions can be targeted toward parts of the water vending supply chain where efficiency gains are most feasible and competition is most vulnerable. However, our research has not uncovered problems that would appear to require urgent attention.

Indeed, water vendors in the Kathmandu Valley have played a crucial role in filling the water supply and demand gap that arose first historically, then critically in recent years, from inadequate public water supply. Approximately 20% of the water for households and business use in the Kathmandu Valley is delivered by this water vending supply chain. However, these water vending services have not been cheap. In the dry season, end users of vended water (both tanker truck water and bottled water) pay approximately 3.4 times as much for vended water as they pay for water from the public piped water distribution system.

These large amounts of money flowing through the water vending supply chain illustrate the revenue potential associated with improved services offered by the public piped distribution system. In Kathmandu, at the time of our writing (2018), the Melamchi Water Supply Project is nearing completion. There is the possibility that KUKL will be able to improve the quantity and quality of the water supplied through the piped water system. The results of our research suggest that most households are already paying much more to water vendors than to KUKL, and that the completion of the Melamchi Water Supply Project offers KUKL a rare opportunity to raise water tariffs simultaneously with the delivery of improved piped services.

Acknowledgements

We thank Professors Bal Kumar KC, Bhim Suwal, and Yogendra Gurung for their research guidance and support in Kathmandu, Nepal, and Dustin Garrick for comments on a previous draft. We also thank the Institute for Water Policy at the National University of Singapore, Lee Kwan Yew School of Public Policy for their financial and institutional support.

Supplementary material

The Supplementary Material for this paper is available online at http://dx.doi.org/10.2166/wp.2019.181.

References

Adams, R. B., Almeida, H. & Ferreira, D. (2009). Understanding the relationship between founder-CEOs and firm performance. *Journal of Empirical Finance 16*, 136–150.

Agrawal, P. & Hall, S. C. (2014). Using accounting metrics as performance measures to assess the impact of information technology outsourcing on manufacturing and service firms. *Journal of Applied Business Research 30*(5), 1559–1568.

Ahlers, R., Schwartz, K. & Perez Guida, V. (2013). The myth of 'Healthy' competition in the water sector: the case of small scale water providers. *Habitat International 38*, 175–182.

Andres, C. (2008). Large shareholders and firm performance – an empirical examination of founding-family ownership. *Journal of Corporate Finance 14*, 431–445.

Appiah Obeng, P., Dwamena-Boateng, P. & Ntiamoah-Asare, D. J. (2010). Alternative drinking water supply in low-income urban settlements using tankers: a quality assessment in Cape Coast, Ghana. *Management of Environmental Quality: An International Journal 21*(4), 494–504.

Asian Development Bank (ADB) (2010). *Project Preparatory Technical Assistance. Consultants' Report for Kathmandu Valley Water Supply and Wastewater System Improvement Study*. Asian Development Bank, Manila.

Babbie, E. (1990). *Survey Research Methods*. Wadsworth, Stamford, CT.

Banerjee, S. G. & Morella, E. (2011). *Africa's Water and Sanitation Infrastructure: Access, Affordability, and Alternatives*. World Bank Publications. Available at: http://documents.worldbank.org/curated/en/712211468202191672/pdf/608040PUB0Afri10-Box358332B01PUBLIC1.pdf.

Bhatia, R. & Falkenmark, M. (1993). *Water Resource Policies and the Urban Poor: Innovative Approaches and Policy Imperatives*. World Bank Publications. Available at: http://www-wds.worldbank.org/external/default/WDSContentServer/WDSP/IB/2008/12/16/000333037_20081216232839/Rendered/PDF/468770WSP0Box31Jan019920global19306.pdf.

Casey, J. F., Kahn, J. R. & Rivas, A. (2006). Willingness to pay for improved water service in Manaus, Amazonas, Brazil. *Ecological Economics 58*(2), 365–372.

Central Bureau of Statistics (CBS) (2012). *National Population and Housing Census 2011 (National Report)*. Available at: http://unstats.un.org/unsd/demographic/sources/census/wphc/Nepal/Nepal-Census-2011-Vol1.pdf.

Collignon, B. & Vézina, M. (2000). *Independent Water and Sanitation Providers in African Cities*. World Bank, Water and Sanitation Program, Washington, DC. Available at: http://nl.ircwash.org/sites/default/files/202.6-00IN-18938.pdf.

Conan, H. & Paniagua, M. (2003). *The Role of Small Scale Private Water Providers in Serving the Poor. Mimeographed Document*. Asian Development Bank, Manila.

Crane, R. (1994). Water markets, market reform and the urban poor: results from Jakarta, Indonesia. *World Development 22*(1), 71–83.

Dixit, A. & Upadhya, M. (2005). *Augmenting Groundwater in Kathmandu Valley: Challenges and Possibilities*. Nepal Water Conservation Foundation, Kathmandu, Nepal. Available at: http://www.academia.edu/download/30977140/R8169-NepalpaperJan05final.pdf.

Gurung, Y., Zhao, J., Bal Kumar, KC, Wu, X., Suwal, B. & Whittington, D. (2017). The costs of delay in infrastructure investments: a comparison of 2001 and 2014 household water supply coping costs in the Kathmandu Valley, Nepal. *Water Resource Research 53*, 7078–7102. doi:10.1002/2016WR019529.

Hunton, J. E., Lippincott, B. & Reck, J. L. (2003). Enterprise resource planning systems: comparing firm performance of adopters and nonadopters. *International Journal of Accounting Information Systems 4*, 165–184.

Janakarajan, S. & Moench, M. (2006). Are wells a potential threat to farmers' well-being? Case of deteriorating groundwater irrigation in Tamil Nadu. *Economic and Political Weekly 41*(37), 3977–3987.

Kathmandu Upatyaka Khanepani Limited (KUKL) (2014). Present Situation of Water Distribution in Kathmandu Valley, Presentation in Kathmandu.

Keener, S., Luengo, M. & Banerjee, S. (2009). *Provision of Water to the Poor in Africa*. Policy Research Working Paper 5387. Available at: http://www.infrastructureafrica.org/system/files/WP13_Standpost.pdf.

Kjellén, M. & McGranahan, G. (2006). *Informal Water Vendors and the Urban Poor*. International Institute for Environment and Development London. Available at: http://www.mumbaidp24seven.in/reference/informal_water_vendors.pdf.

Moench, Y. (2001). *Water Dynamics in a Three Part System: Investigation of the Municipal, Private and Traditional Public Water Supply Systems in Kathmandu, Nepal*. Institute of Social and Environmental Transition and Nepal Water Conservation Foundation, Kathmandu, Nepal.

Opryszko, M. C., Huang, H., Soderlund, K. & Schwab, K. J. (2009). Data gaps in evidence-based research on small water enterprises in developing countries. *Journal of Water and Health 7*(4), 609–622.

Pandey, V. P., Babel, M. S. & Kazama, F. (2009). Analysis of a Nepalese water resources system: stress, adaptive capacity and vulnerability. *Water Science and Technology: Water Supply 9*(2), 213–222.

Pandey, V. P., Babel, M. S., Shrestha, S. & Kazama, F. (2010a). Vulnerability of freshwater resources in large and medium Nepalese river basins to environmental change. *Water Science and Technology 61*(6), 1525–1534.

Pandey, V. P., Chapagain, S. K. & Kazama, F. (2010b). Evaluation of groundwater environment of Kathmandu valley. *Environmental Earth Sciences 60*(6), 1329–1342. doi:10.1007/s12665-009-0263-6.

Pattanayak, S., Yang, J.-C., Whittington, D. & Kumar, B. (2005). Coping with unreliable public water supplies: averting expenditures by Households in Kathmandu, Nepal. *Water Resources Research 41*, W02012. doi:10.1029/2003WR002443.

Shrestha, S., Pradhananga, D. & Pandey, V. P. (2012). *Kathmandu Valley Groundwater Outlook*. Asian Institute of Technology (AIT), Small Earth Nepal (SEN), Centre for Research of Environment Energy and Water (CREEW), International Research Centre for River Basin Environment–University of Yamanashi (ICRE- UY), Kathmandu, Nepal.

SILT Consultants and Development Research and Training Center (1999). *Consumer Survey for Project on Urban Water Supply and Sanitation Rehabilitation*. Technical Report, Kathmandu.

Smith, M. A., Mitra, S. & Narasimhan, S. (1998). Information systems outsourcing: a study of pre-event firm characteristics. *Journal of Management Information Systems 15*(2), 61–93.

Snell, S. (1998). *Water and Sanitation Services for the Urban Poor, Small-Scale Providers: Typology and Profiles*. UNDP–World Bank Water and Sanitation Program Working Paper Series. UNDP, Washington, DC.

Solo, T. M. (1999). Small-scale entrepreneurs in the urban water and sanitation market. *Environment and Urbanization 11*(1), 117–132.

Stickney, C. P., Weil, R. L., Schipper, K. & Francis, J. (2010). *Financial Accounting: An Introduction to Concepts, Methods, and Uses*, 13th edn. South-Western Cengage Learning, Mason, OH.

UN-HABITAT (2007). *Mapping the Footprints of Water Movements in Patan*. UN-HABITAT Water for Asian Cities Programme Nepal, Kathmandu.

Whittington, D., Lauria, D. T., Choe, K., Hughes, J. A., Swarna, V. & Wright, A. (1993). Household sanitation in Kumasi, Ghana: a description of current practices, attitudes, and perceptions. *World Development 21*(5), 733–748.

Whittington, D., Lauria, D. T. & Mu, X. (1991). A study of water vending and willingness to pay for water in Onitsha, Nigeria. *World Development 19*(2), 179–198.

Whittington, D., Lauria, D. T., Okun, D. A. & Mu, X. (1989). Water vending activities in developing countries: a case study of Ukunda, Kenya. *International Journal of Water Resources Development 5*(3), 158–168.

Wutich, A., Beresford, M. & Carvajal, C. (2016). Can informal water vendors deliver on the promise of a human right to water? Results from Cochabamba, Bolivia. *World Development 79*, 14–24.

Zaroff, B. & Okun, D. A. (1984). Water vending in developing countries. *Aqua: Journal of the International Water Supply Association 5*, 289–295.

Zuin, V., Ortolano, L. & Davis, J. (2014). The entrepreneurship myth of small-scale service provision: water resale in Maputo, Mozambique. *Journal of Water, Sanitation and Hygiene for Development 4*(2), 281–292.

Water and time use: evidence from Kathmandu, Nepal

Yvonne Jie Chen[a,*], Namrata Chindarkar[a] and Jane Zhao[b]

[a]*Lee Kuan Yew School of Public Policy, National University of Singapore, Singapore*
**Corresponding author. E-mail: sppcj@nus.edu.sg*
[b]*Department of City and Regional Planning, University of North Carolina at Chapel Hill, Chapel Hill, NC, USA*

Abstract

In this paper, we examine the effect of private tap water reliability on time spent on water collection and total water consumption among urban households in Kathmandu, Nepal. Although the majority of households in Kathmandu are connected to a private tap, they experience intermittent water supply. We link a unique time diary dataset collected between 2014 and 2015 to household water consumption and tap water reliability data. Our empirical analyses demonstrate that improved reliability of private tap water connection (PWC), measured as self-reported reliability and an objective measure of 'probability of getting tap water in the next hour', leads to increased time spent on water collection. Households with more reliable PWC also consume more water overall and from their own taps. Further investigation demonstrates that when private taps became more reliable, households substituted water collected from outside the household, such as water from public taps and public wells, with water from their own private taps. Our results proved robust to additional specification checks.

Keywords: Nepal; Tap water; Urban

1. Introduction

Access to water is costly in many developing countries. Households in these countries spend time and effort on collecting water, often on a daily basis, resulting in productivity and welfare losses. On average, the persons responsible for water collection (usually women and girls) spend 36 min per trip for water collection activities in rural Sub-Saharan Africa and approximately 23 min per trip in rural Asia (Cheung, 2010). In water-scarce regions in India, a woman could spend up to 3 out of her 15 productive hours per day collecting water (Sijbesma *et al.*, 2009).

As a result, many developing countries have implemented large-scale programmes to provide tap water connections to households. The motivations underlying these programmes include the provision of safe and convenient water supply and consequently improve household welfare. However, rapid expansion of tap water connections can lead to inadequate supply if it is not accompanied by

doi: 10.2166/wp.2019.082

improvements to the supporting water infrastructure, thus undermining the effectiveness of providing tap water connections.

In this paper, we study the effect of reliability of private tap water connections (PWCs) on household time allocation and household welfare measured as volume of water consumed in Kathmandu, Nepal[1]. The importance of infrastructure on development and growth has been demonstrated in the existing literature. However, the importance of infrastructure quality has not received adequate attention, due to the thinking that quality will eventually catch up. In reality, in many countries, it takes decades, if not longer, for the infrastructure to operate at optimal capacity. During this lengthy period of catching up, households usually rely on existing sources of water supply, such as public taps or wells and private water markets to manage their household water supply. An interesting question that arises is whether access to tap water, even though unreliable, is better than nothing?

Tap water systems provide treated water from public sources to end users. The provision of tap water requires supporting infrastructure, such as water pipelines that link end users directly to the water source. However, as in many developing countries, the tap water system in Kathmandu suffers from intermittent service and poor reliability. Although over 70% of households had access to tap water in 2001, water pressure and running time varied significantly, owing mainly to weak supporting water infrastructure, such as aging and broken pipes, and depleting water sources (Whittington *et al.*, 2002; Pattanayak *et al.*, 2005). A majority of households get tap water only for a few hours every day. Under these conditions, it is unclear what effect improved reliability would have on water collection times. At the time of the writing of this article, we were unaware of any other study examining time use patterns under conditions of unreliable water supply.

We use unique household data collected in the Kathmandu Valley for our analyses. Kathmandu Valley is the economic and political centre of Nepal. With its rapid urbanisation and growing population, Kathmandu Valley has experienced significant water shortages. Although in our sample, 87% of households had access to tap water, the connections can be highly unreliable. Data on household water usage collected by the authors from 2014 to 2015 indicate that, on average, households with tap water connectivity received water from their tap for only 9 days out of a month, and for the days when water was available, it only runs for less than 100 min. Tap water reliability contributes significantly to household time allocation decisions. During our survey, many respondents suggested that they often needed to leave work early in order to turn on the tap when the water became available.

We provide a simple theoretical model that illustrates the effect of increased productivity of household water production (that is, when tap water is more reliable) on households' time use and water consumption. Our model shows that without imposing specific functional forms on households' water production, the effects remain ambiguous. If a linear production function is imposed, households would be expected to consume more water but the amount of time they spend on collecting water still remains ambiguous.

We use a subsample of 819 households connected to tap water for our empirical analyses. Using the detailed household survey and time diary, we first demonstrate that the time spent on water collection activities is negatively correlated with productive activities. For every one minute increase in time spent on water collection, there is about half a minute reduction in time spent on work. However, we find no evidence that time spent on water collection affects the probability of employment. We also demonstrate that the reliability of tap water was positively correlated with time spent on water collection. In particular, if a household had regular tap water, water collectors spend 17–23 min more on water collection per

---

[1] For simplicity, private tap water connections are referred to as 'tap water' hereafter.

day (compared to a household with less regular tap water). Further analysis suggested that this increase was largely driven by the increase in time spent on collecting water from the private tap. In estimating the relationship between tap water reliability and water consumption, we find the reliability measure to be positively correlated with total water consumption and water consumed from the tap. In the meantime, households with reliable tap water consume less water from public sources (such as public wells and public taps) and water vendors. Note that the empirical findings we present in this paper can only be interpreted as correlations and not causal relationships owing to potential endogeneity in access to tap water as well as its reliability. We perform additional robustness checks to confirm if our estimates were sensitive to the choice of control variables or different fixed effects.

Our study adds to the current literature on reliable access to water and time use in a number of ways. First, this study contributes to the understanding of tap water reliability and time use in urban areas of a less developed country. As women are usually responsible for water collection in developing countries, most of the existing literature has tended to focus on the effects of improved access to water on female labour markets and domestic labour supply outcomes (Ilahi & Grimard, 2000; Koolwal & van de Walle, 2013). Little is known, empirically, about how access to tap water affects time spent on water collection and other daily activities[2]. Second, we provide new evidence that can be added to the existing literature on infrastructure reliability and intermittent water supply. Past literature that evaluated the effect of infrastructure tended to focus on a categorical change, and less attention has been paid to its quality and reliability (Chen *et al.*, 2019). Our results indicate that intermittent water supply could hurt labour productivity if water becomes available during working hours. It also suggests that there might be a welfare effect of improved reliability of tap water. Households with more reliable tap water spend less time outside the house collecting water and consume more from the tap. Therefore, although the total amount of time they spend on collecting water is more, they can probably accomplish other things alongside water collection at home, leading to a potential welfare gain. This study is also the first to use a detailed time diary to estimate the effect of reliable tap water on time use. It allows us to disaggregate time allocation across broad activity categories.

2. Conceptual framework

In this section, we present a simple theoretical model that links the productivity of household water production with time use and water consumption. Suppose a household is set out to maximise $U(c_1; c_2; l)$, where c_1 is the consumption of water, c_2 is the consumption of other market goods and l is leisure. Suppose water can be collected from a private tap or other sources (such as a private well or a public tap). The household's problem can be written as follows:

$$
\begin{aligned}
&\mathrm{Max}_{t1,t2,tw} U(c_1, c_2, l) \\
&\text{s.t.:}\quad c_1 = g(t_1;\ \theta_1) + v(t_2;\ \theta_2) \\
&pc_2 = wt_w \\
&t_1 + t_2 + t_w + l = T
\end{aligned} \tag{1}
$$

[2] Studies on this topic in the existing literature have usually focused on rural households. For example, Gross *et al.* (2018) demonstrated that the installation of public water access points significantly reduced time spent on water collection in rural Benin. Meeks (2017) discovered that improved water access in rural Kyrgyzstan increased the time spent on market activities and leisure.

where $g(t_1;\ \theta_1)$ is the production function of water from a private tap and $v(t_2;\ \theta_2)$ is the production function of water from other sources. θ_1 and θ_2 are parameters in the production functions, which indicate the productivity of water production from various sources. p is the price for market goods. Household is endowed with time T, which can be allocated to the production of water from the private tap (t_1) or other sources (t_2) to the labour market at wage rate w (t_w) and leisure (l).

Note that Equation (1) is equivalent to maximising $U(g(t_1) + v(t_2),\ wt_w/p,\ T - t_1 - t_2)$. The research question of interest is as follows: What is the effect of θ_1 on time allocation and water consumption? We can show that these effects are ambiguous, that is, the signs of the partial derivatives $\partial t_1/\partial\theta_1$, $\partial t_w/\partial\theta_1$ and $\partial c_1/\partial\theta_1$ cannot be determined[3]. Intuitively, this is because as the productivity of tap water production increases, there are two forces driving time t_1 allocated to this activity. On the one hand, the substitution effect will lead to an increase in t_1 compared to other activities, as the production of tap water becomes more efficient in terms of labour (hence per unit price becomes cheaper). However, this substitution effect can be bounded above as the quantity of water consumed in a household cannot increase infinitely. On the other hand, an increase in tap water productivity will have an income effect. This income effect will push demand for all consumption upwards, including demand for market goods and leisure. This income effect will induce households to supply more labour to the labour market in exchange for market goods. In equilibrium which effect dominates is ambiguous.

Koolwal & van de Walle (2013) discussed a special case of the model presented in Equation (1), where the tap water production function is assumed to be linear, that is, $g(t_1) = \rho t_1$. Under this assumption, if a household reallocates time spent on producing water from a private tap, t_1, to the labour market, it can earn a wage income of wt_1. If the price of a market good is assumed to be numerarie, the reallocation of time results in wt_1 units of additional consumption of the market good, but now the household produces t_1 units of water from its private tap. Therefore, the relative price of water produced from the private tap to market good is w/ρ. Applying the Slutsky equation, it can be shown that $\partial c_1/\partial\rho$ is always negative. Therefore, the authors conclude that under the assumption of a linear production function, as ρ increases, the consumption of home-produced water will always increase. Nonetheless, even under this restrictive assumption, the sign of $\partial t_1/\partial\rho$ remains ambiguous.

To summarise, when tap water becomes more reliable, it is unclear from the model whether time spent on water collection will increase or decrease. This lack of clarity is due to the fact that the substitution effect and income effect from such a change work in opposite directions. However, if stricter assumptions on the water production function are imposed, an increase in total water consumption should be expected while the effect on time allocation remains ambiguous.

3. Data and outcome variables

Data for the empirical analyses presented in this paper were drawn from a household survey of 1,500 households in five municipalities in the Kathmandu Valley in 2014 and 2015 (Kathmandu, Lalitpur, Bhaktapur, Kirtipur and Madhyapur). This was a follow-up survey of the 2001 wave in the same area. The two waves were merged to construct a panel. Details of the sampling strategy and data for the 2001 wave can be found in Whittington *et al.* (2002) and Pattanayak *et al.* (2005), and details of the 2014–2015 survey can be found in Yogendra *et al.* (2017). In this paper, we only use data from the 2014–2015 wave. The survey

[3] The proofs are omitted here.

was conducted using a multi-stage clustered sampling method, and the sample is representative of the population of Kathmandu in 2001 but not of the population in 2014–2015. The 2001 survey collected information on household characteristics and water usage as well as perceptions of tariffs and willingness-to-pay. In 2014–2015, an attempt was made to re-survey all the households in the 2001 sample to construct a panel. If a household could not be located in the 2014–2015 survey, the current resident or the immediate neighbour was interviewed. In total, 61.8% of households were successfully re-surveyed. We only use households that were surveyed in both rounds for our analyses.

The 2014–2015 survey included a time diary module. The time diary was distributed at the time of survey, and in it, household members were asked to reflect on the activities that they performed in a typical day over a 24-hour period, using 30 min as the smallest block (see Figure 1). Following the standards of the established literature, we provided 26 types of activities for respondents to choose from Devoto *et al.* (2012). The respondent of the main survey, spouse of the respondent, and the person most responsible for water collection (if different from respondent or spouse) filled out the time diary. We grouped the 26 activities into seven broad categories: work, leisure, household chores, social activities, rest, water collection and meal times.

Literature on survey methods and cognitive psychology has highlighted potential bias that might be introduced in time use data arising from recall and salience of activities (Menon, 1993; Beegle *et al.*, 2012; Arthi *et al.*, 2018). In particular, the concern is that respondents are likely to

3. TIME-USE SCHEDULE (Imagine a normal 24-hour day)
Respondent: (1) Respondent from first part survey OR (2) his/her spouse
About Whom: (1) Respondent from first part of the survey (2) Spouse of the respondent (3) Person most responsible for collecting water for the household (if different from the respondent or spouse) and (4) children aged 7-15 years who mostly assists in collecting water for the household

301. Respondent's name: ________________ .... ID (From HH Roster) ☐☐

302. Who are other responsible for collecting water in your household? (Check the boxes by √ that applies)

1. Respondent................ ☐ 2. Spouse of respondent ...... ☐ 3. Other family member (if different)..... ☐ 4. Children (7-15 yrs) .... ☐

303. Among them, who is the most responsible person for collecting water? **Name:** ________________ ......ID (From HH Roster): ☐☐

304. TIME	305. CODE				304. TIME	305. CODE			
	Ask for the boxes in Q302 are checked					Ask for the boxes in Q302 are checked			
	Respondent	Spouse	Other	Children		Respondent	Spouse	Other	Children
ID					ID				
5:00 - 5:30					17:00 - 17:30				
5:30 - 6:00					17:30 - 18:00				
6:00 - 6:30					18:00 - 18:30				
6:30 - 7:00					18:30 - 19:00				
7:00 - 7:30					19:00 - 19:30				
7:30 - 8:00					19:30 - 20:00				
8:00 - 8:30					20:00 - 20:30				
8:30 - 9:00					20:30 - 21:00				
9:00 - 9:30					21:00 - 21:30				
9:30 - 10:00					21:30 - 22:00				
10:00 - 10:30					22:00 - 22:30				
10:30 - 11:00					22:30 - 23:00				
11:00 - 11:30					23:00 - 23:30				
11:30 - 12:00					23:30 - 24:00				
12:00 - 12:30					24:00 - 00:30				
12:30 - 13:00					0:30 - 1:00				
13:00 - 13:30					1:00 - 1:30				
13:30 - 14:00					1:30 - 2:00				
14:00 - 14:30					2:00 - 2:30				
14:30 - 15:00					2:30 - 3:00				
15:00 - 15:30					3:00 - 3:30				
15:30 - 16:00					3:30 - 4:00				
16:00 - 16:30					4:00 - 4:30				
16:30 - 17:00					4:30 - 5:00				

SCHEDULE CODES	
1	Sleep
2	Bathing and freshening
3	Tea time/breakfast
4	Read newspaper
5	Dressing, getting ready
6	Commuting to/from work/school/college
7	At work (include self-employment/own business)
8	Eat meals
9	Cooking
10	Washing and cleaning
11	Collecting water
12	Other domestic chores
13	Resting
14	Go to the market
15	Take care/help children and aged in the family
16	Spend time with persons in the household
17	Socialize with friends, neighbors, community
18	Professional training
19	Religious and spiritual activities
20	Entertainment – TV, DVDs, radio, internet, movies
21	Phone calls, letters, emails
22	Exercise – walking, jogging, gym, sports
23	At school/college, doing homework, studying
24	Playing (indoors or outdoors)
25	Pursue hobbies – reading, writing, drawing, music, dance
26	Other (Specify)

Fig. 1. Time use survey (time diary) for the 2014–2015 survey.

over(under)-report time spent on different activities depending on the recall period and whether the activities are salient and regular. For instance, if time spent on collecting water is less salient but regular, such as 'every afternoon', then respondents may not respond with precise recall-and-count strategy but instead just recall the periodicity (Menon, 1993).

We designed the time use module acknowledging these limitations. First, we did not impose a specific recall period; instead, we asked the respondents about a 'typical day'. Second, we did not explicitly highlight certain activities, such as 'how much time do you spend on collecting water'. Instead, we let the respondents recall their daily activities chronologically over a 24-hour period, and then we post-coded those activities. It is possible that our survey data collection method may have introduced bias such as respondents sub-consciously reporting more time spent on collecting water since they were informed *a priori* that the survey was about water supply. However, taking into consideration the resource and time constraints of implementing the survey, this was the best way forward.

Similarly, the data we constructed for the quantity of water consumed from various sources are subject to measurement errors. Existing literature has usually relied on self-reported water usage or imputed data from recalled water expenditure to infer household water consumption volumes (Fuente *et al.*, 2016; Apoorva *et al.*, 2018). The difficulties in accurately estimating the quantity of water consumed are twofold. First, confidentiality concerns prevent disclosure of socioeconomic status in water bills. As a result, researchers are not able to link metered water usage to household characteristics in many countries. In addition, households in developing countries usually gather water from various sources, many of which are not metered. Water consumption from these sources can only be estimated based on recall data. In our survey, we also relied on the recall method to gather information on the amount of water consumed. We asked households about their water consumption from 11 sources. For consumption from sources other than private taps, we follow the standard practices of the existing literature and asked respondents to report the size of their containers and the number of containers they retrieved per trip (Apoorva *et al.*, 2018).

The two sets of outcome variables of interest are time spent on collecting water and total water consumption from various sources. We discuss the main variables used in the empirical analyses in the following paragraphs.

3.1. Time spent on water collection

We constructed two variables to measure time spent on water collection. The first variable is 'activity_water'. This variable was constructed from the time diary module. Respondents only reported the amount of time they spent on collecting water (if at all) during the day and did not differentiate it by source. Therefore, we could not disaggregate time spent on collecting water from a private tap or from other sources. To supplement this, we constructed a second variable 'activity_water_outside' using information from the household survey. This variable measures the time (minutes per day on average) spent on collecting water from a public tap, public well or a stone tap. Time spent on collecting water from outside the household was computed by adding up round trip time plus waiting time for collection from these sources. This number was reported by the main survey respondent, and he or she could be different from the person who was most responsible for water collection activities in the household. Our data indicated that out of the 1,500 respondents for the main household survey, 412 filled out the time diary module (out of a total of 819 time diaries). Hence, the variable activity_water_outside is subject to measurement errors, and the results need to be interpreted with caution.

3.2. Water consumption from various sources

We collected data for water consumption from the following sources: a private tap, any public sources (including public tap, public well and stone tap), a private well, water vendors and total water consumption from all sources. These variables were constructed from the main household survey conducted in the 2014–2015 period. In particular, in the main household survey, each respondent (usually the head of the household who may not be the person most responsible for water collection) was questioned about the availability of various water sources. We constructed the water consumption variables by adding up consumption in the corresponding categories.

3.3. Measures of productivity of home water production

We constructed two measures of reliability for tap water connection. The first measure is a qualitative, ordinal measure of reliability. The respondent of the household survey was asked to judge the reliability of tap water on a scale of 1–4, with 1 being 'very regular' and 4 being 'unreliable', for both the dry and rainy seasons. We generated a dummy variable that is equal to 1 if the respondent reports that the tap water connection is 'very regular' or 'regular'. We used the responses for reliability of tap water in the dry season in our main results.

The second reliability measure uses Hashimoto *et al.* (1982)'s definition of a system in water resources planning: the frequency or probability that a system is in a satisfactory state. We calculated the probability of getting tap water in the next hour. Specifically, we computed the following measure:

$$\text{reliable\_hrs} = \frac{\text{tap\_minutes}/60}{24 \times (1 + \text{tap\_gap})}$$

tap_minutes are the number of minutes of tap water a household expects to receive every time they receive water from the tap. tap_gap is the number of days between the time tap water is received. Therefore, the measure reliable_hrs computes the probability that a household receives tap water in the next hour. For example, if a household receives uninterrupted tap water supply 24 hours a day, 7 days a week, the household's probability of getting tap water in the next hour would be $((24 * 60)/60)/(24 * (1 + 0)) = 100\%$.

Summary statistics for the sample used in the analysis are presented in Table 1. Since only the person most responsible for water collection answered the time diary, there were some missing observations if this person were absent or unavailable at the time of the survey. Out of the time diaries distributed to 1,500 households in the original sample, 819 of them were collected. We compared household income and household size of the 819 households against those that were missing from the original sample. Our t-test indicated that although the logarithm of household income was not statistically different between the two groups (P-value = 0.82 for two-sided test), households that answered the time diary survey had a slightly smaller household size (difference = 0.2, P-value = 0.037). We attempted to address the sample selection issue by including household controls and individual characteristics. We also performed additional robustness checks which are discussed in Section 5.

As shown in Table 1, although all households are connected to a private tap in our sample, on average, the probability for a household to receive tap water in the next hour is extremely low at only 2%. Households waited an average of 5 days for tap water to become available; these periods of availability

Table 1. Summary statistics.

Variable	Description	Observed	Mean	Standard deviation	Minimum	Maximum
Key variables						
Time spent on water collection						
Activity water	Time spent on collecting water per day (minutes)	819	43.7	41.5	0.00	240
Activity water outside	Time spent on collecting water outside household per day (minutes)	819	8.70	34.1	0.00	540
Water consumption – dry season						
Water total dry	Total water consumption per day in the dry season (litre)	819	301	362	6.70	2,775
Water pipe dry	Total water consumption per day from pipe in the dry season (litre)	819	78.6	136	0.00	2,000
Water public dry	Total water consumption per day from public well, tap or stone tap in the dry season (litre)	819	23.4	51.3	0.00	500
Water priwell dry	Total water consumption per day from private well (priwell) in the dry season (litre)	819	136	318	0.00	2,500
Water vendor dry	Total water consumption per day from vendors in the dry season (litre)	819	55.0	139	0.00	1,667
Water consumption – rainy season						
Water total rain	Total water consumption per day in the rainy season (litre)	819	361	413	5.00	2,917
Water pipe rain	Total water consumption per day from pipe in the rainy season (litre)	819	160	269	0.00	2,500
Water public rain	Total water consumption per day from public well, tap or stone tap in the rainy season (litre)	819	25.5	57.5	0.00	500
Water priwell rain	Total water consumption per day from private well in the rainy season (litre)	819	133	308	0.00	2,000
Water vendor rain	Total water consumption per day from vendors in the rainy season (litre)	819	26.7	82.6	0.00	800
Reliability measures						
Reliable dry	Self-reported reliability of tap water connection in the dry season	819	0.24	0.43	0.00	1.00
Reliable rain	Self-reported reliability of tap water connection in the rainy season	819	0.75	0.44	0.00	1.00
Reliable hours	Probability of getting tap water in the next 24 h	575	1.56	1.89	0.06	20.8
Control variables						
hhsize	Household size	819	4.96	2.05	1.00	15.0
Age	Age of the respondent	819	46.3	14.0	13.0	90.0
Gender	Gender of the respondent	819	1.77	0.42	1.00	2.00
Education	Whether respondent has received middle school education	818	0.65	0.48	0.00	1.00
Inschool	Whether respondent is currently enrolled in school	819	1.97	0.18	1.00	2.00
Married	Whether the respondent is married	819	0.81	0.39	0.00	1.00
Lnincome	Logarithm of monthly household income (Rs)	795	10.4	1.28	6.21	16.1
Other time use variables (minutes per day)						
Work	Time spent at work	819	131	198	0.00	750
School/training	Time spent at school	819	7.22	50.0	0.00	570
Bathing/dressing	Time spent on taking a bath or dressing	819	36.7	19.3	0.00	150
Meals	Time spent on eating meals	819	136	44.8	0.00	270
Commute	Time spent on commuting	819	10.5	31.4	0.00	600
Sleep	Time spent on sleeping	819	441	50.4	270	630
Chores	Time spent on running household chores other than water-related activities	819	313	183	0.00	810
Leisure	Time spent on leisure activities	819	319	177	0.00	870

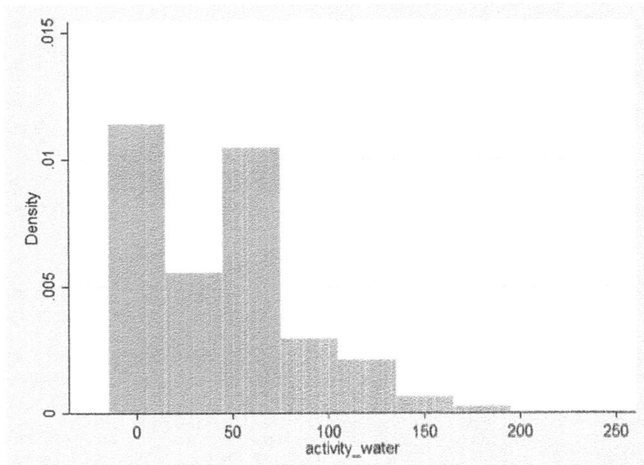

Fig. 2. Distribution of time spent on water collection. *Note*: This figure shows the histogram of water collection activities (minutes/day) among the 819 respondents that completed the time diary in the 2014–2015 survey.

on average lasted for approximately 100 min. A typical respondent still spent about 40 min on water collection activities every day. This included all water-related activities, including fetching water from the tap and from other resources within and outside the household. This number suggests that although households were 'connected' to a private tap, they still spent a great deal of time on collecting water. Figure 2 plots the histogram of time spent on water collection among the 819 households. The figure suggests that there was significant variation in time spent on collecting water among these households, and one possible explanation for this is the reliability of tap water connections across these households.

4. Descriptive analyses of tap water reliability, time use and water consumption

In this section, we present descriptive analyses of tap water reliability and time use. In particular, we examine the time use patterns across different household activities for households with good and poor connections.

4.1. Water collectors

We paired the time diary records of the person most responsible for collecting water for the household with information in the household roster in the main household survey. Therefore, we had detailed data about who these water collectors were. Table 2 demonstrates that 43.8% of main water collectors were spouses and 28.9% were heads of households. It can be seen that 76.8% of water collectors were female. Daughters-in-law were also frequently tasked with collecting water. We also observed that sons, daughters, parents and other family members participated in collecting water. While the majority of water collectors were female, male water collectors made up 23% of the sample. Figure 3 shows the distribution of the age of water collectors. They had a mean (and median) age of 46. Only 11 households (out of 819) reported water collectors under the age of 18. We saw no significant participation of

Table 2. Water collector's household role.

	Male	Female	Total
Spouse	1	358	359
Head	143	94	237
Son/daughter-in-law	1	121	122
Son/daughter	39	34	73
Father/mother	0	6	6
Other non-relative	2	4	6
Brother/sister-in-law	0	5	5
Grandchild	2	2	4
Brother/sister	0	3	3
Other relative	1	1	2
Nephew/niece	1	0	1
Father/mother-in-law	0	1	1
Total	190	629	819

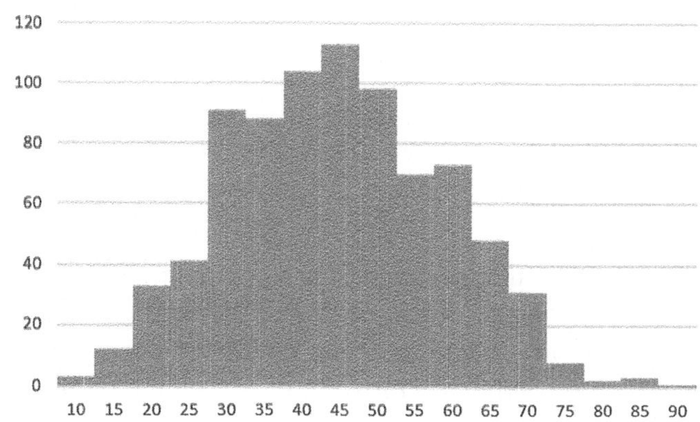

Fig. 3. Age distribution of water collectors.

children in the collection of water. Our results, therefore, differed from much of the trend observed in the literature, as we have a significant portion of male collectors and few collectors who were children.

Additionally, households often included more than one member responsible for collecting water. The results revealed that 82% of households had one main water collector, 18% had two collectors and 0.4% had three. Of the main collectors, one-third (33%, $N = 276$) reported working in a typical day.

4.2. Water collection and work

Next, we examined the water collection time patterns throughout the day. Figure 4 demonstrates the water collection patterns for the entire sample and contrasts it with working hours. Each bar represents the number of households reporting water collection during that half hour. It can be observed that a large portion of the water collection (in blue) occurs in the morning before 10:00 am. There is also a

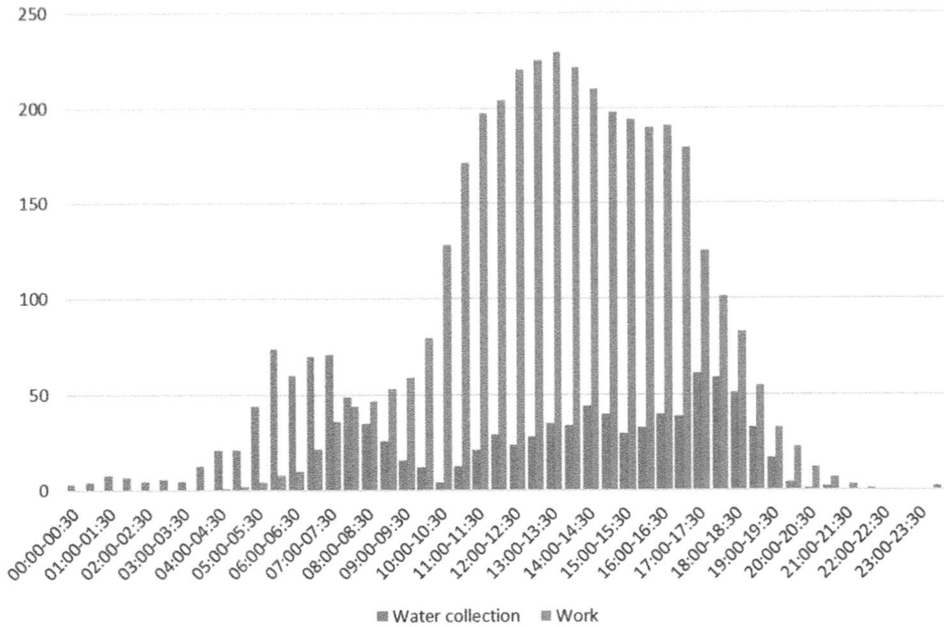

Fig. 4. Water collection time. *Note*: This figure displays frequencies of time intervals for water collection and work for all 819 respondents that completed the time diary in the 2014–2015 survey. Please refer to the online version of this paper to see this figure in colour: http://dx.doi.org/10.2166/wp.2019.082.

significant portion of water collection that occurs during work hours (illustrated using orange bars) and in the evening. Two smaller peaks can be observed at 2:00 pm and 5:00 pm.

Because much of the literature has tended to focus on labour supply outcomes, we examined the relationship between water collection and work. First, we examined the observable socioeconomic differences between water collectors who work more and those who work less. We run regressions of the number of hours worked in a typical day on socioeconomic variables and on hours spent on collecting water and water reliability. In Table 3, it can be observed that in regressions (1) and (2), which controls only for socioeconomic characteristics, females, older people, individuals from bigger households and those currently enrolled in school work fewer hours. Marital status and education had no significant impact on hours worked. Fixed effects for municipality or ward or both were included in all models. Regressions (3)–(8) build upon models (1) and (2) by including variables related to tap water reliability and hours spent on collecting water. There was a significant negative coefficient on time spent on collecting water – more time spent on collecting water was associated with less time spent working. Reliability did not have a significant effect on hours worked. Table 4 presents the effect of time spent on water collection on the probability of employment. The coefficients on activity_water were significant when municipality fixed effects were controlled for but became insignificant when ward fixed effects were controlled for instead. This trend suggests that the observed correlation between water collection activities and the probability of employment might be driven by unobserved ward characteristics.

We investigated the time trade-off between work and water collection more closely by looking at the subset of participants who both worked and collected water in a typical day. Figure 5 shows the patterns of work and water collection only for the subset of those who both worked and collected water ($N = 167$).

Table 3. Regressions of time spent on water collection on hours worked.

	Dependent variable = activity prod							
	(1)	(2)	(3)	(4)	(5)	(6)	(7)	(8)
Activity water			−0.773***	−0.497***	−0.789***	−0.517***	−0.727***	−0.440**
			(0.156)	(0.172)	(0.160)	(0.173)	(0.189)	(0.204)
Reliable dry			2.371	−1.326				
			(15.698)	(17.651)				
Reliable rain					7.180	11.496		
					(15.814)	(16.268)		
Reliable hours							0.827	0.122
							(4.429)	(5.085)
Lnincome	12.053***	11.605***	12.413***	11.844***	12.334***	11.817***	13.788***	14.174***
	(2.553)	(2.580)	(2.492)	(2.533)	(2.499)	(2.535)	(3.262)	(3.390)
Household size	−13.909***	−14.050***	−13.553***	−13.461***	−13.456***	−13.387***	−13.888***	−15.060***
	(3.115)	(3.095)	(3.015)	(3.049)	(3.020)	(3.048)	(3.700)	(3.553)
Age	−3.014***	−3.051***	−3.164***	−3.112***	−3.149***	−3.103***	−3.425***	−3.338***
	(0.483)	(0.481)	(0.483)	(0.480)	(0.482)	(0.481)	(0.587)	(0.598)
Gender	−150.921***	−148.143***	−158.654***	−151.675***	−159.009***	−152.037***	−171.595***	−167.033***
	(18.790)	(18.318)	(18.445)	(18.191)	(18.425)	(18.168)	(22.020)	(21.981)
Education	−2.727	−0.658	3.797	2.764	4.367	3.734	5.376	0.712
	(14.146)	(14.266)	(14.000)	(14.259)	(14.126)	(14.378)	(16.726)	(16.913)
In school	130.100***	112.464***	132.422***	112.067***	133.105***	113.871***	151.471***	137.258**
	(42.344)	(42.440)	(42.779)	(42.808)	(42.953)	(43.108)	(51.691)	(54.015)
Married	−24.381	−18.112	−26.140	−19.753	−25.904	−19.947	−32.121	−25.213
	(18.467)	(18.563)	(18.320)	(18.555)	(18.260)	(18.530)	(22.241)	(22.319)
Observations	790	790	790	790	790	790	556	556
R^2	0.137	0.192	0.161	0.200	0.161	0.200	0.171	0.219
Municipal FE	Y	N	Y	N	Y	N	Y	N
Ward FE	N	Y	N	Y	N	Y	N	Y

Notes: ***, ** and * denote significance level at 1%, 5% and 10%, respectively. Robust standard errors are in parentheses. Activities are measured in minutes/day. See Section 3 for more details on variable definitions. Y, yes; N, no.

Table 4. Regressions of time spent on water collection on the probability of employment.

	Dependent variable = employed							
	(1)	(2)	(3)	(4)	(5)	(6)	(7)	(8)
Activity water			−0.001***	−0.001	−0.001***	−0.001	−0.001***	−0.001
			(0.000)	(0.000)	(0.000)	(0.000)	(0.000)	(0.001)
Reliable dry			0.044	0.032				
			(0.039)	(0.043)				
Reliable rain					0.006	0.004		
					(0.038)	(0.039)		
Reliable hours							0.010	0.012
							(0.010)	(0.012)
Lnincome	0.031***	0.029***	0.031***	0.029***	0.031***	0.029***	0.034***	0.033***
	(0.007)	(0.007)	(0.007)	(0.007)	(0.007)	(0.007)	(0.009)	(0.009)
Household size	−0.034***	−0.034***	−0.033***	−0.033***	−0.033***	−0.033***	−0.027***	−0.030***
	(0.008)	(0.008)	(0.008)	(0.008)	(0.008)	(0.008)	(0.010)	(0.009)
Age	−0.008***	−0.008***	−0.008***	−0.008***	−0.008***	−0.008***	−0.008***	−0.008***
	(0.001)	(0.001)	(0.001)	(0.001)	(0.001)	(0.001)	(0.002)	(0.002)
Gender	−0.334***	−0.322***	−0.346***	−0.326***	−0.346***	−0.326***	−0.372***	−0.356***
	(0.039)	(0.039)	(0.039)	(0.039)	(0.039)	(0.039)	(0.047)	(0.047)
Education	−0.016	−0.005	−0.005	0.000	−0.005	−0.001	0.034	0.029
	(0.034)	(0.035)	(0.034)	(0.035)	(0.034)	(0.035)	(0.040)	(0.041)
In school	0.430***	0.381***	0.439***	0.383***	0.434***	0.381***	0.450***	0.404***
	(0.107)	(0.110)	(0.107)	(0.110)	(0.107)	(0.111)	(0.134)	(0.141)
Married	−0.034	−0.010	−0.039	−0.015	−0.036	−0.012	−0.044	−0.024
	(0.044)	(0.045)	(0.044)	(0.045)	(0.044)	(0.045)	(0.052)	(0.053)
Observations	790	790	790	790	790	790	556	556
R^2	0.137	0.176	0.148	0.179	0.147	0.178	0.157	0.186
Municipal FE	Y	N	Y	N	Y	N	Y	N
Ward FE	N	Y	N	Y	N	Y	N	Y

Notes: ***, ** and * denote significance level at 1%, 5% and 10%, respectively. Robust standard errors are in parentheses. Activities are measured in minutes/day. See Section 3 for more details on variable definitions. Employed = 1 if the individual has a valid occupation code in the household roster and is not a student/retiree/housemaker. Y, yes; N, no.

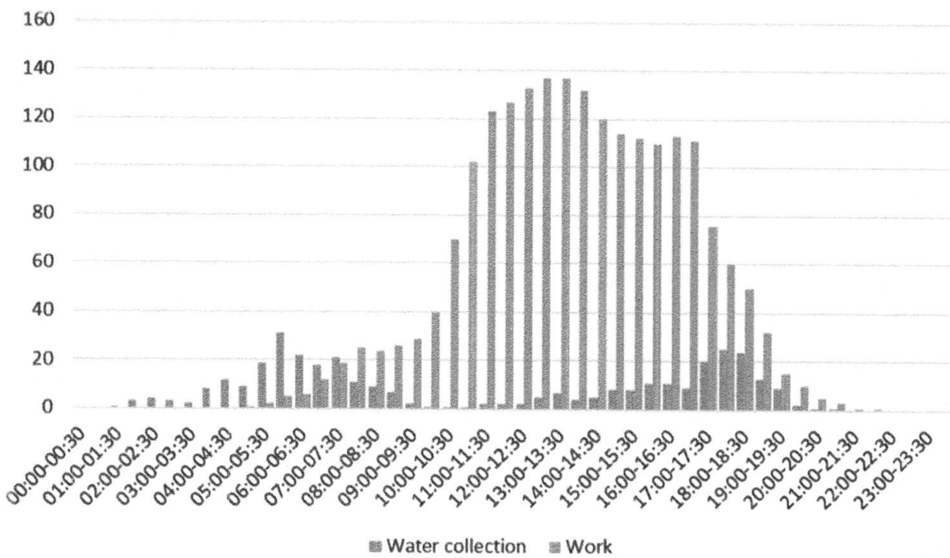

Fig. 5. Water collection time for workers. *Note*: This figure displays frequencies of time intervals for water collection and work for 167 (out of all 819) respondents that both worked and were the person most responsible for water collection.

Of these 167 water collectors, 26 worked right before water collection, four worked right after water collection and seven worked both before and after water collection. It is for these 37 water collectors (22% of this subsample) that time spent on collecting water was likely to directly affect the amount of time spent working. These were people who collected water in the afternoon, as shown in Figure 4 as the areas where water collection and work overlap. Of those who both worked and collected water, 41% worked in retail, 17% in other industries, 14% in government positions and 11% in wage labour. Out of 167 water collectors, 93% ($N = 155$) were employed (which meant that they were not in school or retired). In addition, almost none of these people (less than 2%) worked as wage labourers. Water collectors are also less likely to be employed. Out of the 819 water collectors, two-thirds reported not having a full time job (no occupation code was identified), while the proportion was only one-third of the entire sample. Hence, estimates presented in Tables 3 and 4 might be subject to selection bias as people who self-select themselves for water collection duties might have differed from the general population.

An ideal analysis would have estimated the correlation between time spent on water collection and the productive time of all household members. However, due to the restrictions of the survey, we were unable to perform this analysis. Alternatively, in Table 5, we estimate the correlation between time spent on water collection and the fraction of household members that were employed. We discovered no correlation between the two. In summary, our results suggest that for those water collectors that do work, the time they spent on productive activities were negatively correlated with time spent on water collection. However, there is limited evidence that time spent on water collection affects employment decisions.

4.3. Water collection and other activities

In our time diaries, we not only asked respondents to recall the time they spent on water and work but also collected information on other activities performed throughout the day. Therefore, we had the

Table 5. Regressions of time spent on water collection on the percentage of household members employment.

	Dependent variable = % adult household members that are employed							
	(1)	(2)	(3)	(4)	(5)	(6)	(7)	(8)
Activity water			−0.000	0.000	−0.000	0.000	−0.000	0.000
			(0.000)	(0.000)	(0.000)	(0.000)	(0.000)	(0.000)
Reliable dry			0.021	0.031				
			(0.020)	(0.022)				
Reliable rain					0.007	0.010		
					(0.018)	(0.019)		
Reliable hours							−0.006	−0.005
							(0.004)	(0.004)
Lnincome	0.027***	0.025***	0.027***	0.025***	0.026***	0.025***	0.029***	0.028***
	(0.005)	(0.006)	(0.005)	(0.006)	(0.005)	(0.006)	(0.006)	(0.007)
Household size	−0.016***	−0.016***	−0.016***	−0.016***	−0.016***	−0.016***	−0.017***	−0.018***
	(0.004)	(0.004)	(0.004)	(0.004)	(0.004)	(0.004)	(0.005)	(0.005)
Age	0.001	0.001	0.001	0.001	0.001	0.001	0.001	0.001
	(0.001)	(0.001)	(0.001)	(0.001)	(0.001)	(0.001)	(0.001)	(0.001)
Gender	−0.014	−0.007	−0.015	−0.005	−0.015	−0.005	−0.011	−0.004
	(0.020)	(0.020)	(0.020)	(0.020)	(0.021)	(0.020)	(0.024)	(0.024)
Education	−0.025	−0.020	−0.024	−0.021	−0.023	−0.021	−0.024	−0.023
	(0.016)	(0.017)	(0.016)	(0.017)	(0.016)	(0.017)	(0.020)	(0.020)
In school	0.065	0.037	0.068	0.040	0.066	0.039	0.060	0.049
	(0.054)	(0.055)	(0.055)	(0.056)	(0.054)	(0.055)	(0.070)	(0.067)
Married	−0.058**	− 0.044*	−0.060**	− 0.046*	−0.058**	− 0.043*	−0.049	− 0.033
	(0.024)	(0.025)	(0.024)	(0.025)	(0.024)	(0.025)	(0.031)	(0.032)
	(0.042)		(0.043)		(0.042)		(0.057)	
Observations	790	790	790	790	790	790	556	556
R^2	0.092	0.124	0.094	0.127	0.093	0.125	0.097	0.139
Municipal FE	Y	N	Y	N	Y	N	Y	N
Ward FE	N	Y	N	Y	N	Y	N	Y

Notes: ***, ** and * denote significance level at 1%, 5% and 10%, respectively. Robust standard errors are in parentheses. Activities are measured in minutes/day. See Section 3 for more details on variable definitions. A dependent variable is computed as the number of household members employed/ household size. Y, yes; N, no.

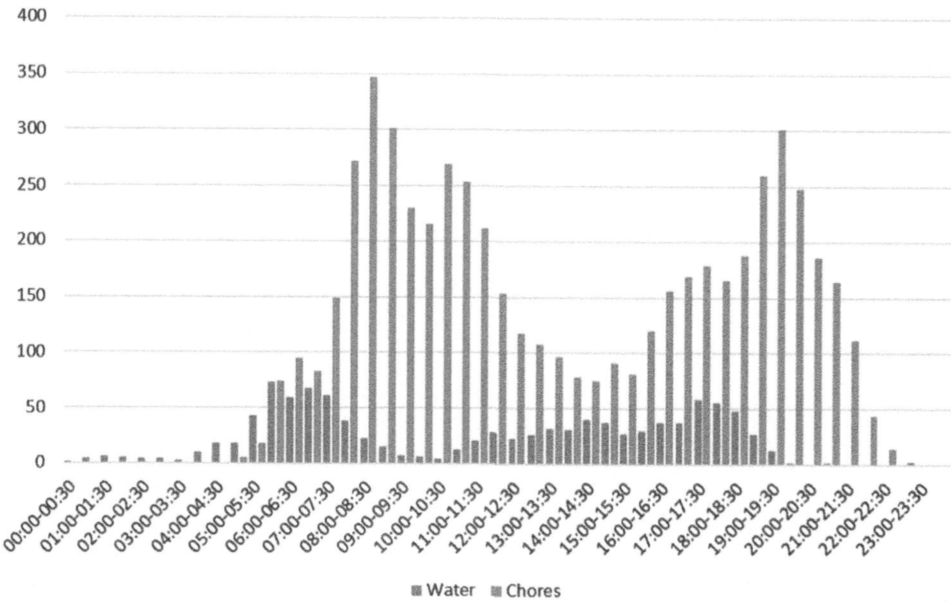

Fig. 6. Chores and water collection. *Note*: This figure displays frequencies of time intervals for water collection and non-water chores for all 819 respondents that completed the time diary in the 2014–2015 survey.

opportunity to examine the relationship between tap water reliability and a wide range of activities, including non-water collection chores, leisure, eating meals, having tea, bathing and sleeping.

First, we examined the distribution of time spent on non-water collection chores and water collection (Figure 6). Water collection tended to occur in the morning before all other chores. Most of the chores occurred in the morning, between breakfast and lunch, when water collection activities were also low. Some water collection took place in the afternoon and increased at approximately noon, 2:00 pm and the late afternoon. Thereafter, another large portion of the chores was completed, from late afternoon through the evening. Trade-offs between water collection and other chores did not seem likely.

Figure 7 shows the time distributions for leisure and water collection. We observed significant overlaps between peak water collection times and peak leisure times, which make direct trade-offs between leisure and water collection more likely.

4.4. Aggregate daily time use patterns

This section of the analysis separates the households into four categories based on combinations of: if they have a private well, a PWC, and if the PWC is reliable or not. Figure 8 illustrates the proportions of the subsample that participate in each activity type during each half hour. These figures provide an idea of which activities are occurring during which times, on aggregate.

Figure 8 shows that daily time use patterns are broadly similar across the four types of households. It can be observed that water collection usually occurs in the morning or in the late afternoon. If the person most responsible for water collection was not collecting water in the morning, she or he spent time sleeping, bathing, eating meals and doing leisure activities. If water collection did not happen in the afternoon, then the person most responsible for water collection spent time reaching the end of their

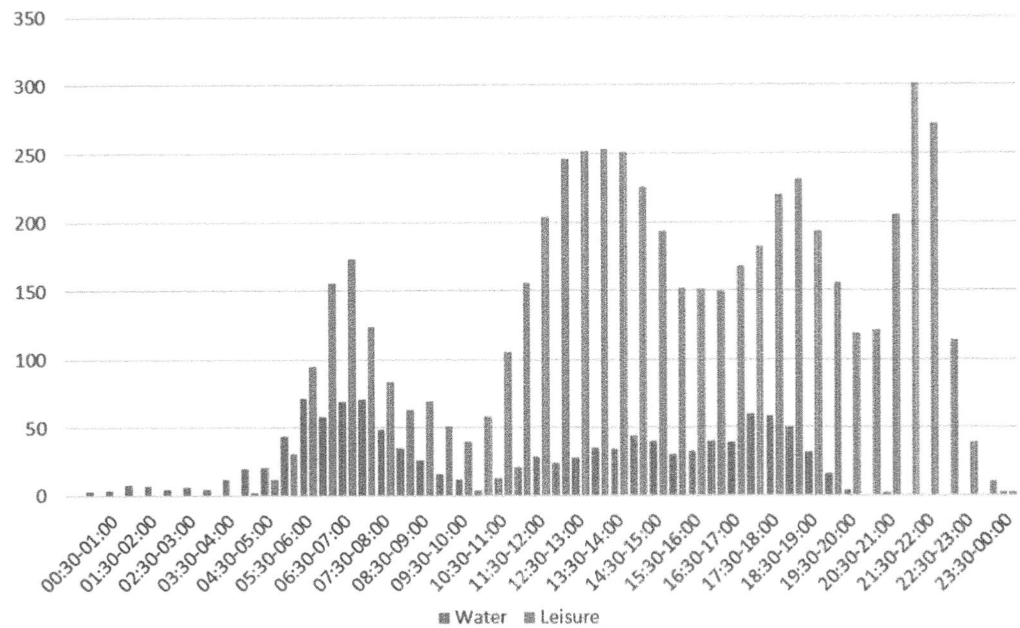

Fig. 7. Leisure and water collection. *Note*: This figure displays frequencies of time intervals for water collection and leisure for all 819 respondents that completed the time diary in the 2014–2015 survey.

work day, working on other chores or engaging in leisure activities. Right before and after water collection in the evening, many households ate meals, worked on other chores and engaged in leisure activities.

According to these results, it appears that water collectors in households with reliable connections spend more time on collecting water and less time working than do water collectors in households with unreliable connections. This is confirmed with average hours spent (Table 6). It shows that, interestingly, water collectors in households with unreliable connections work more than water collectors from the other subsamples (148.6 and 121.34 min per day vs. 122.14 and 115 min per day). They also spent less time on chores (315.32 and 295.17 min per day vs. 357.32 and 304.64 min per day). Households with unreliable water connections also slept more (450.13 and 439.83 min per day vs. 429.38 and 423.14 min per day). At the same time, we observed more water collectors who attend school or other training in the reliable subsamples than in the unreliable subsamples. Households with both a tap water connection and a private well spent the most amount of time on collecting water (59 min per day). Water collectors from these households also worked the least and spent more time eating meals. Water collectors from households without either a reliable tap water connection or a private well spent the least time on water collection and school/training and spent the most time sleeping and commuting compared to other household types.

Theories outlined in Section 2 provide explanation for these seemingly counterintuitive results. Regression analyses in Section 5 also confirm the descriptive findings presented here.

4.5. Summary of findings

From the results, we draw three conclusions. First, households' daily time use patterns are very similar. Second, while water collection time patterns closely reflect leisure and chore time patterns, the

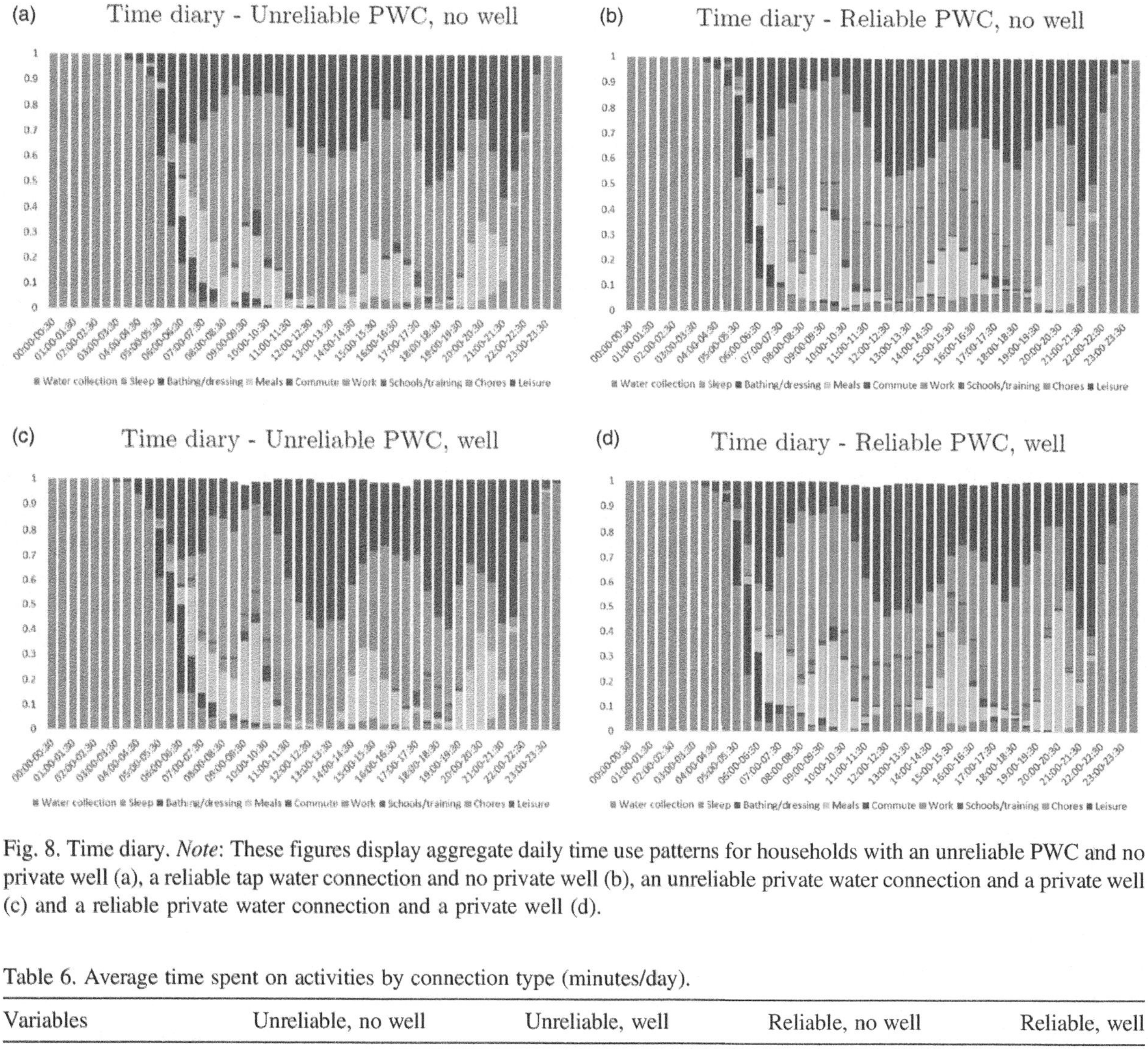

Fig. 8. Time diary. *Note*: These figures display aggregate daily time use patterns for households with an unreliable PWC and no private well (a), a reliable tap water connection and no private well (b), an unreliable private water connection and a private well (c) and a reliable private water connection and a private well (d).

Table 6. Average time spent on activities by connection type (minutes/day).

Variables	Unreliable, no well	Unreliable, well	Reliable, no well	Reliable, well
Activity water	37.91	36.88	56.25	59.64
Work	148.60	121.34	122.14	115.00
School/training	2.55	5.96	5.09	11.07
Bathing/dressing	34.72	37.40	34.82	37.50
Meals	128.17	138.29	135.80	146.79
Commute	11.23	10.17	4.55	9.29
Sleep	450.13	439.83	429.38	432.14
Chores	315.32	295.17	357.32	304.64
Leisure	309.32	350.96	294.64	322.50
Observed	235	292	112	84

overlap with work time in the afternoons is concerning from a policy perspective. Finally, while households with unreliable tap water connections spend the least time on collecting water, these households also purchase much more water from private vendors (tankers and jar water), suggesting that households substitute tap water with vended water.

5. Regression analyses of tap water reliability, time use and water consumption

5.1. Empirical specification

In this section, we use ordinary least squares (OLS) regressions to estimate the correlation between time spent on water collection, water consumption and reliability of tap water in more detail. The baseline OLS regression is as follows:

$$y_{\mathrm{iwm}} = \alpha_0 + \alpha_1 \,\mathrm{Reliability}_{\mathrm{iwm}} + \alpha_2 X_{\mathrm{iwm}} + \alpha_3 \tau_{\mathrm{m}} + \alpha_4 \sigma_{\mathrm{w}} + u_{\mathrm{iwm}} \qquad (2)$$

where y_{iwm} are the two sets of outcomes on time allocation and water consumption as described in Section 3. $\mathrm{Reliability}_{\mathrm{iwm}}$ is the self-reported tap water reliability measure, which takes value 1 if it is regular and 0 otherwise, or the probability of getting tap water in the next hour. For household *i*, in ward *w*, and municipality *m*, X_{iwm} is a set of individual and household characteristics as observed in 2014–2015, which includes gender, education, age, age-squared, log of household income and its quadratic form and household size. We also include municipality or ward fixed effects (or both) to capture any municipality-specific or ward-specific characteristics that would affect time spent on water collection.

5.2. Results

Table 7 presents the OLS results of tap water reliability on time spent on collecting water and time spent on collecting water outside the household. Recall that while activity_water is reported by the person most responsible for water collection, activity_water_outside is reported by the main survey respondent. In panels A, B and C, we report the effect of three reliability measures, that is, the reliability of tap water in the dry season, in the rainy season and the probability of getting tap water in the next hour. Results in panel A indicate that greater reliability of tap water in the dry season is significantly correlated with more time spent on water collection and less time spent on water collection activities outside the household. Results in panel B demonstrate that greater reliability of tap water in the rainy season increases total time spent on water collection but is uncorrelated with time spent on water collection outside the household. Results in panel C suggest that an increase in the probability of getting tap water in the next hour is positively correlated with time spent on water collection activities and negatively correlated with water collection outside the household.

In summary, results in Table 7 suggest that more reliable tap water is positively correlated with time spent on water collection. There is also some evidence that this increase is mainly driven by time spent on collecting water from a private tap within the household.

Table 8 shows the effect of tap water reliability on water consumption. Similar to the previous table, we group results into four panels with respect to the three reliability measures. Panels A and B report the effect of reliability in the dry season and the rainy season on water consumption in dry and rainy seasons, respectively. Panels C and D report the effect of 'reliable_hrs' on water consumption in dry and rainy seasons, respectively. Column (1) shows a positive and significant correlation between reliability

Table 7. Effect of reliability on time use – OLS.

Variables	(1) Activity water	(2) Activity water	(3) Activity water outside	(4) Activity water outside
Panel A				
Reliable dry	16.812*** (3.411)	7.405** (3.742)	−6.607** (2.831)	−3.421 (2.379)
N	766	766	766	766
R^2	0.107	0.284	0.151	0.102
Panel B				
Reliable rain	22.824*** (2.811)	12.160*** (2.909)	1.766 (1.663)	3.004 (2.375)
N	766	766	766	766
R^2	0.135	0.293	0.144	0.102
Panel C				
Reliable hours	1.905* (0.980)	1.113 (1.033)	−0.562*** (0.179)	−0.624** (0.273)
N	766	766	766	766
R^2	0.107	0.284	0.151	0.102
Municipal FE	Y	N	Y	N
Ward FE	N	Y	N	Y
Household controls	Y	Y	Y	Y
Individual controls	Y	Y	Y	Y

Notes: ***, ** and * denote significance level at 1%, 5% and 10%, respectively. Robust standard errors are in parentheses. Each cell reports the coefficient from a separate regression. Activities are measured in minutes/day. See Section 3 for more details on variable definitions. All regressions control for log household income, household size, individual age, gender, education, whether in school and marital status. Y, yes; N, no.

and total water consumption for three out of the four reliability measures when municipality fixed effects are included. In column (2), the probability of getting water in the next hour was still positively correlated with total water consumption when ward fixed effects are controlled for. The magnitude of coefficients in panel C suggests that a 1 percentage point increase in the likelihood of receiving water in the next hour leads to 47.5 more litres of total water consumption per day. Columns (3)–(10) present the results for each water source. As columns (3) and (4) indicate, unsurprisingly, reliability of tap water is associated with more water consumption from the private tap. These results are significant with or without the ward fixed effects. Results in columns (5)–(10) suggest substitution of consumption between water received from the tap and that from other sources. Furthermore, tap water reliability is negatively associated with water consumed from a private well, public tap (or stone tap) or from water vendors.

6. Robustness checks

We conducted two robustness checks to confirm whether our results were sensitive to the choice of control variables and sample selection. Results are presented in Tables 9 and 10.

Table 9 presents our results after we re-run our OLS regressions by adding more household control variables, including age square, survey month and gender of the household head. Each cell in the table is a separate regression, and we report coefficients for reliable_dry, reliable_rain and reliable_hrs in panels A, B, C and D. Columns (1)–(4) of panel D are intentionally left blank as these coefficients are the same as columns (1)–(4) in panel C. Columns (5)–(8) in panel C estimate the effect of 'reliable_hrs' on water consumption in the dry season. Columns (5)–(8) in panel D estimate its effect on water consumption in

Table 8. Effect of reliability on water consumption – OLS.

Variables	(1) Water total	(2) Water total	(3) Water pipe	(4) Water pipe	(5) Water public	(6) Water public	(7) Water priwell	(8) Water priwell	(9) Water vendor	(10) Water vendor
Panel A										
Reliable dry	60.538**	−22.641	122.607***	79.474***	−17.716***	−3.624	−10.956	−69.339***	−29.138***	−25.849*
	(27.713)	(31.040)	(17.206)	(17.274)	(4.090)	(3.947)	(22.885)	(25.294)	(9.503)	(13.585)
N	766	766	766	766	766	766	766	766	766	766
R^2	0.128	0.273	0.179	0.238	0.196	0.233	0.101	0.300	0.051	0.089
Panel B										
Reliable rain	42.567	−39.219	120.961***	64.537***	2.753	8.116*	−47.430	−79.885***	−29.205***	−33.001***
	(33.630)	(28.466)	(16.342)	(17.727)	(3.657)	(4.285)	(29.677)	(24.434)	(7.902)	(8.430)
N	766	766	766	766	766	766	766	766	766	766
R^2	0.129	0.298	0.117	0.223	0.184	0.285	0.095	0.296	0.068	0.099
Panel C										
Reliable hours (dry)	36.603**	34.918**	27.647***	22.379**	−1.662***	−1.140*	8.625	11.757	2.661	2.411
	(14.709)	(14.440)	(10.335)	(10.063)	(0.510)	(0.597)	(8.671)	(7.962)	(6.752)	(6.167)
N	542	542	542	542	542	542	542	542	542	542
R^2	0.130	0.320	0.154	0.341	0.125	0.315	0.090	0.300	0.047	0.088
Panel D										
Reliable hours (rain)	66.104***	57.141***	67.267***	53.079***	−1.856***	−1.127**	3.812	7.617	−1.736	−1.860
	(20.077)	(19.236)	(17.872)	(16.490)	(0.493)	(0.535)	(7.214)	(6.607)	(2.632)	(2.575)
N	542	542	542	542	542	542	542	542	542	542
R^2	0.171	0.392	0.235	0.406	0.156	0.408	0.076	0.301	0.066	0.090
Municipal FE	Y	N	Y	N	Y	N	Y	N	Y	N
Ward FE	N	Y	N	Y	N	Y	N	Y	N	Y
Household controls	Y	Y	Y	Y	Y	Y	Y	Y	Y	Y
Individual controls	Y	Y	Y	Y	Y	Y	Y	Y	Y	Y

Notes: ***, ** and * denote significance level at 1%, 5% and 10%, respectively. Robust standard errors are in parentheses. Each cell reports the coefficient from a separate regression. Activities are measured in minutes/day. See Section 3 for more details on variable definitions. Panel C estimates the effect of reliable hours on dry season water consumption. Panel D estimates the effect of reliable hours on rainy season water consumption. All regressions control for log household income, household size, individual age, gender, education, whether in school and marital status. Y, yes; N, no.

Table 9. Robustness checks – adding more control variables.

Dependent variables	(1) Activity water	(2) Activity water	(3) Activity water outside	(4) Activity water outside	(5) Water total	(6) Water total	(7) Water pipe	(8) Water pipe
Panel A								
Reliable dry	16.593***	6.136*	−7.398***	−4.046**	62.930**	−27.188	122.006***	79.622***
	(3.258)	(3.685)	(2.433)	(1.970)	(26.617)	(30.470)	(15.880)	(16.205)
N	819	819	819	819	819	819	819	819
R^2	0.085	0.255	0.102	0.068	0.072	0.228	0.156	0.208
Panel B								
Reliable rain	23.260***	12.176***	2.966* (1.754)	5.229* (2.866)	38.742	−50.418*	119.722***	60.933***
	(2.679)	(2.818)			(32.969)	(28.565)	(16.112)	(16.832)
N	819	819	819	819	819	819	819	819
R^2	0.117	0.266	0.096	0.070	0.084	0.255	0.096	0.190
Panel C								
Reliable hours (dry)	1.980**	1.068 (1.006)	−0.504***	−0.558**	42.231***	38.723**	28.821***	23.089**
	(0.979)		(0.164)	(0.249)	(15.737)	(15.121)	(10.309)	(9.927)
N	575	575	575	575	575	575	575	575
R^2	0.099	0.294	0.108	0.086	0.084	0.279	0.134	0.344
Panel D								
Reliable hours (rain)					72.435***	61.018***	70.629***	55.251***
					(20.847)	(19.688)	(18.064)	(16.522)
N					575	575	575	575
R^2					0.141	0.363	0.226	0.406
Municipal FE	Y	N	Y	N	Y	N	Y	N
Ward FE	N	Y	N	Y	N	Y	N	Y
Household and individual controls	Y	Y	Y	Y	Y	Y	Y	Y
Additional controls	Y	Y	Y	Y	Y	Y	Y	Y

Notes: ***, ** and * denote significance level at 1%, 5% and 10%, respectively. Robust standard errors are in parentheses. Each cell reports the coefficient from a separate regression. Activities are measured in minutes/day. See Section 3 for more details on variable definitions. All regressions control for log household income, household size, individual age, gender, education, whether in school and marital status. In addition, we also control for age square, gender of head of household and survey month.

Columns (1)–(4) of panel D are left blank purposely as these coefficients are the same as columns (1)–(4) in panel C. Columns (5)–(8) in panel C estimate the effect of reliable hours on dry season water consumption. Columns (5)–(8) in panel D estimate the effect of reliable hours on rainy season water consumption. Y, yes; N, no.

Table 10. Robustness checks – drop observations surveyed in August.

Dependent variables	(1) Activity water	(2) Activity water	(3) Activity water outside	(4) Activity water outside	(5) Water total	(6) Water total	(7) Water pipe	(8) Water pipe
Panel A								
Reliable dry	18.442***	6.806	−2.746 (1.960)	−3.167* (1.855)	34.699	0.064	119.274***	85.893***
	(3.932)	(4.562)			(28.386)	(33.337)	(21.009)	(22.447)
N	556	556	556	556	556	556	556	556
R^2	0.115	0.329	0.218	0.103	0.083	0.216	0.180	0.222
Panel B								
Reliable hours (dry)	3.176**	1.099	−0.658***	−0.311**	63.703***	50.154**	39.407***	33.286**
	(1.412)	(1.564)	(0.183)	(0.143)	(19.813)	(20.165)	(12.862)	(14.070)
N	387	387	387	387	387	387	387	387
R^2	0.145	0.360	0.121	0.157	0.158	0.277	0.222	0.378
Municipal FE	Y	N	Y	N	Y	N	Y	N
Ward FE	N	Y	N	Y	N	Y	N	Y
Household and individual controls	Y	Y	Y	Y	Y	Y	Y	Y
Additional controls	Y	Y	Y	Y	Y	Y	Y	Y

Notes: ***, ** and * denote significance level at 1%, 5% and 10%, respectively. Robust standard errors are in parentheses. Each cell reports the coefficient from a separate regression. Activities are measured in minutes/day. See Section 3 for more details on variable definitions. We only report the effect of reliability in dry seasons in this table as all samples included in this table are surveyed in the dry season. All regressions control for log household income, household size, individual age, gender, education, whether in school and marital status. Y, yes; N, no.

the rainy season. We reported both results with municipal fixed effects or ward fixed effects. As observed, with these additional controls, our main results were largely consistent other than the effect of reliable_rain on total water consumption in column (6). When ward fixed effect was controlled for, the effect was negative and statistically significant. A possible explanation is that households without reliable tap water might store extra water or over-consume during the rainy season when water is relatively cheap, while consumption decisions are less affected for households with reliable tap water.

Table 10 presents the results when households surveyed in August are excluded. We performed this analysis in August as it is typically the last month of the rainy season in the Kathmandu Valley. By retaining results for dry season water consumption and dry season reliability, we sought to reduce potential recall errors due to the timing of survey. We ran the regressions again using the main specification, and the results were similar.

7. Conclusion

In this paper, we analysed the relationship between household time allocation, reliability of tap water and water consumption patterns under conditions of intermittent supply. We discover that time spent on productive activities is negatively correlated with time spent on water collection for the person most responsible for water collection in the household. We also demonstrate that when tap water connections become more reliable, households spend more time on collecting water. As a consequence of this time investment, households consume more water both from their private tap and overall.

From a policy perspective, because an increase in reliability leads to an increase in collection time inside the home, it is important to arrange water supply schedules in a way that minimises disruptions that can affect work, leisure and other activities. For cities with intermittent supply, more research is needed to understand how individuals and households prioritise different activities in order to identify which times of day would be the most conducive to water collection. Additionally, since we were unable to record simultaneous activities within the same time block, it may be relevant to determine what other activities occur while the tap is turned on and the water collector is waiting for the storage containers to be filled. Finally, with households with unreliable water connections shifting away from collecting water outside the home to vended water, there is room to examine the trade-offs between collecting and buying water.

References

Apoorva, R., Biswas, D. & Srinivasan, V. (2018). Do household surveys estimate tap water use accurately? Evidence from pressure-sensor based estimates in Coimbatore, India. *Journal of Water, Sanitation and Hygiene for Development 8*(2), 278–289.

Arthi, V., Beegle, K., De Weerdt, J. & Palacios-Lopez, A. (2018). Not your average job: measuring farm labor in Tanzania. *Journal of Development Economics 130*, 160–172.

Beegle, K., Carletto, C. & Himelein, K. (2012). Reliability of recall in agricultural data. *Journal of Development Economics 98*(1), 34–41.

Chen, Y. J., Chindarkar, N. & Xiao, Y. (2019). The effect of reliable electricity on health facilities, health information, and health services utilization: evidence from rural Gujarat, India. *Journal of Health, Population and Nutrition 38*(1), 7.

Cheung, P. (2010). *The World's Women 2010 Trends and Statistics*. Department of Economics and Social Affairs, the United Nations, Washington, DC.

Devoto, F., Duo, E., Dupas, P., Pariente, W. & Pons, V. (2012). Happiness on tap: piped water adoption in urban Morocco. *American Economic Journal: Economic Policy 4*(4), 68–99.

Fuente, D., Gatua, J. G., Ikiara, M., Kabubo-Mariara, J., Mwaura, M. & Whittington, D. (2016). Water and sanitation service delivery, pricing, and the poor: an empirical estimate of subsidy incidence in Nairobi, Kenya. *Water Resources Research 52*(6), 4845–4862.

Gross, E., Gunther, I. & Schipper, Y. (2018). Women are walking and waiting for water: the time value of public water supply. *Economic Development and Cultural Change 66*(3), 489–517.

Hashimoto, T., Stedinger, J. R. & Loucks, D. P. (1982). Reliability, resiliency, and vulnerability criteria for water resource system performance evaluation. *Water Resource Research 18*(1), 14–20.

Ilahi, N. & Grimard, F. (2000). Public infrastructure and private costs: water supply and time allocation of women in rural Pakistan. *Economic Development and Cultural Change 49*(1), 45–75.

Koolwal, G. & van de Walle, D. (2013). Access to water, women's work, and child outcomes. *Economic Development and Cultural Change 61*(2), 369–405.

Meeks, R. C. (2017). Water works: the economic impact of water infrastructure. *Journal of Human Resources 52*(4), 1119–1153.

Menon, G. (1993). The effects of accessibility of information in memory on judgments of behavioral frequencies. *Journal of Consumer Research 20*(3), 431–440.

Pattanayak, S. K., Yang, J. C., Whittington, D. & Kumar, K. C. B. (2005). Coping with unreliable public water supplies: averting expenditures by households in Kathmandu, Nepal. *Water Resources Research 41*(2), 1–11.

Sijbesma, C., Verhagen, J., Nanavaty, R. & James, A. J. (2009). Impacts of domestic water supply on gender and income: results from a participatory study in a drought-prone region in Gujarat, India. *Water Policy 11*(1), 95–105.

Whittington, D., Pattanayak, S., Yang, J. C. & Kumar, K. C. B. (2002). Household demand for piped water services in Kathmandu, Nepal. *Water Policy 4*(6), 531–556.

Yogendra, G., Zhao, J., KC, B. K., Wu, X., Suwal, B. & Whittington, D. (2017). The costs of delay in infrastructure investments: a comparison of 2001 and 2014 household water supply coping costs in the Kathmandu Valley, Nepal. *Water Resource Research 53*(8), 7078–7102.

Constructing a fabula of resilience: a lived experience approach

Leong Ching[a,*], Zhou Yishu[a], Nisha Francine Rajoo[b,c] and Sarah Ruiqi Tan[d]

[a]*Institute of Water Policy, Lee Kuan Yew School of Public Policy, National University of Singapore, 469A Bukit Timah Road, #02-03A, Level 2, Tower Block, Singapore 259770, Singapore*

**Corresponding author. E-mail: spplcc@nus.edu.sg*

[b]*Lee Kuan Yew School of Public Policy, National University of Singapore, Singapore*

[c]*Supreme Court of Singapore, Singapore*

[d]*Department of Geography, National University of Singapore, 1 Arts Link, Singapore 117570, Singapore*

Abstract

Recent research in the studies of socio-ecology has raised an intriguing side effect associated with coping strategies and perceptions of environmental risk – where a perceived ability to cope with the effects of climate change and ecological degradation leads to an over-estimation of self-reliance and resilience. Known as the 'paradox of resilience', little has been understood about how or when such over-estimations occur. This paper sheds light on such phenomena by constructing a fabula (or narrative structure) of the paradox by investigating the case of extreme water shortage and the coping strategies of residents in the Kathmandu Valley, Nepal. Despite the long delay in providing public access to water, and the high cost this imposes on the public, members of the public have grown accustomed to the unreliable water supply, and willingly accept the high and coping costs of finding alternative sources of water. Using the lived experiences method, the paper reveals a link between self-perceptions of resilience, the incremental nature of the problem, a perceived ability to pay coping costs, and a lack of confidence in government capacity.

Keywords: Coping; Emotions; Resilience; Water governance

Introduction

Despite the increasing frequency and impact of environmental problems, such as climate change, resource scarcity, and natural disasters, as well as widening consensus that vigorous action is necessary to mitigate the adverse impacts of these problems, the reaction among the general public remains one of relative apathy (Richler, 2017).

doi: 10.2166/wp.2019.083

Instead, many paradoxes and anomalies are inherent in the public's relationship with and attitude toward the environment. Marshall found that oftentimes, victims of flooding, drought, or severe storms become '*less* willing to talk about climate change or even accept that it is real', and that having children can make people *less* concerned about climate change (Marshall, 2014, p. 2).

This does not mean that people are unconcerned or unaware of the threat of climate change. Rather, it belies the 'paradox of social resilience' (Shaw *et al.*, 2014), in which the act of 'coping' gives people what the authors call the 'delusion of resilience', and results in inaction. This is an unintended result of individual or social resilience, whereby people paradoxically take on greater risks than they are able to manage due to the unfounded conviction that they are well equipped to handle it, or resign themselves to absorbing the coping costs involved rather than recognizing it as a situation of crisis. In the context of environmental or ecological issues, this interpretation suggests that people are not deliberately undermining the threats of climate change – rather, individuals are taking on extraneous coping costs due to a systematic miscalculation of the risks involved.

Within extant literature, there are two possible explanations for this. The first is 'emotional numbing' as a socio-emotional coping mechanism. Norgaard (2011) documents a curious reluctance by the Norwegian public to discuss and engage with climate change issues, despite a high level of knowledge about such issues. She suggests that the pessimism and fear surrounding climate change issues have paradoxically created a need for Norwegians to 'live in denial' and continue their lives as usual, avoiding prolonged thought or discussion of the issue, and only divulging their feelings about the environment when pressed.

The same emotional numbing argument has been presented in the work of Eckersley (2008) and Futerra (2005), who argue that fears of a bleak future have the potential to generate maladaptive and nihilistic responses, in contrast to more adaptive activist responses. In the specific case of climate change, dramatic, fearful or shocking images of climate change are 'likely to distance or disengage individuals from climate change, tending to render them feeling helpless and overwhelmed when they try to comprehend their own relationship with the issue' (O'Neill & Nicholson-Cole, 2009:375).

A second explanation for this reaction to environmental scarcity and governmental failures is a rational calculus of trade-offs. Figueiredo *et al.* (2009) show that while indigenous communities in Agueda, Portugal, accurately estimate the risk of local flooding, they continue to occupy liminal spaces in flood plains due to their optimism about other benefits such as employment and livelihoods.

This suggests that cognitive reasoning and decision-making also play a part in regulating emotional responses. Thus, expected negative emotional responses to vulnerability (e.g. fear, anger) must be read in the context of other factors which may promote a positive emotional response, such as job security. However, such conscious trade-offs may, in fact, be an implicit risk-taking strategy, as it is difficult to assure that other benefits (such as agricultural livelihoods) will remain relevant in the long run under unpredictable and uncertain processes such as climate change.

In a similar vein, González-Riancho *et al.* (2015) observe a low-risk perception of storm surge flooding amongst residents of Dithmarschen, a low-lying area along the German North Sea coast with two major storm surge events in the last century, including a 1962 storm surge which claimed 340 lives. Instead, the top two problems perceived to be affecting stakeholders were livelihood-related difficulties and demographic change and migration.

These findings are echoed by Boissière *et al.* (2013) work on local perceptions of climate change in Mamberamo, Papua, Indonesia, where locals considered climate variation to have little impact on their livelihoods when contrasted with other factors, such as anthropogenic changes to the watershed through the logging and mining industries, political decentralization, and infrastructure development. The prioritization of other risks has therefore been thought to be a rational response for communities that appear calm about certain environmental risks.

In this paper, a resilience response is exemplified by populations that display positive adaptation and coping mechanisms despite experiences of significant adversity or trauma caused by social or environmental injustices. This is primarily a psychological conception located within the work of scholars such as Luthar *et al.* (2000), Masten *et al.* (1999), and Rutter (1999, 2000).

In this sense, it departs from traditional hydrological conceptions of resilience, as a function of *systems* as per Holling, which focuses more on empirical systems, such as infrastructure or institutions, and their ability to 'absorb change and disturbance and still maintain the same relationships between populations or state variables' (Holling, 1973:14).

As such, we interpret the outcome of general apathy toward climate change as symptomatic not of a *lack* of concern but rather its presence, by way of the paradox of resilience and people's use of cognitive and emotional arithmetic in order to understand and relate to more complex and abstract subjects. This generates two outcomes, which form the basis of our two hypotheses:

H1: Resilient response due to emotional numbing, resulting from pessimism and fear.
H2: Resilient response due to rational calculus of trade-offs.

H1 and H2 have quite different implications for policy-makers. The first relies on a denial of reality and masks a vulnerability that may be exposed over time. Policy-makers may then have to deal both with present fears while providing a strong safety net in order to pre-empt the manifestation of these fragilities. The second assumes a population able to make reasoned choices and to recalibrate their responses adaptively to suit environmental changes. This would be closer to 'true' resilience. However, the question for government is the extent to which government policies can or should impact such resilient behavior – government interventions could lead to perverse outcomes if the paradox holds. Of course, H1 and H2 are not mutually exclusive, and one line of enquiry in this paper is to examine whether and how they can be integrated.

The case of Kathmandu in Nepal presents a unique case to study this paradox.

The Kathmandu Valley is the largest metropolis and most populous area in Nepal. It faces rapid urbanization, modernization, and population growth. As its urban area expands to accommodate more inhabitants, the lack of attendant proper land-use planning has resulted in the growth of peri-urban sprawl along the fringes of the Valley. The sprawl exacerbates problems of sanitation and water services, causing severe water insecurity and associated health risks.

While the Kathmandu Upatyaka Khanepani Limited (KUKL) – the government's water utility – is ostensibly responsible for the Valley's water resources management, it has largely failed to ensure a safe and secure supply of water for most residents. The lack of reliable water supply has resulted in significant economic coping costs for residents who resort to alternative sources of water supply, including private wells, storage tanks, or purchasing water from tanker trucks and bottled water vendors (Gurung *et al.*, 2017). Therefore, despite their poor socio-economic status, many residents in the Valley still spend a substantial proportion of household income on the basic necessity of potable water (Gurung *et al.*, 2017).

A survey by Guragai *et al.* (2017) evaluating consumers' perceptions of the water supply found that nearly half of the respondents (48%) rated water services as average – suggesting satisfaction, if not acceptance at the very least, of the quality of water provision in the Valley. However, 39% of respondents also rated piped water services as very poor, with only 13% of respondents considering it satisfactory. This shows that although consumers are critical of the quality of water services in the Valley, this does not translate into an impetus for broader collective action or public pressure on the KUKL or local government. Rather, consumer efforts are channeled toward dealing with rising coping costs and searching for supplementary water sources.

Two factors that characterize the residents of the Kathmandu Valley – high coping costs and the lack of participation in decision-making concerning the chronic water shortage – allow us to study the paradox in a pure sense. In other words, we are able to exclude all other hypotheses for low salience (lack of direct experience, lack of knowledge, and lack of certainty about the nature of the problem due to its complexity) (Laerhoven, 2014).

This paper studies the paradox of social resilience by constructing a fabula of this phenomenon. A fabula is defined as a series of logically and chronologically related events that are caused or experienced by characters in a story world, or what Bal calls the 'logic of events' (2009:7).

This construction is motivated by an impulse to understand environmental behavior. It starts from a structuralist premise that 'the series of events presented in story answers to the same rules as those controlling human behavior' (Bal, 2009:7).

The fabula is based on the following conditions:

1. There is direct experience, knowledge, and certainty about the problem.
2. There are coping costs involved.
3. There is a resilient response to the problem.

A fabula is distinct from a 'story' in the sense that a story is an instantiation from a fabula; that is to say, a story is told from a certain point of view. In this paper, for example, a particular story of resilience can be captured from one single respondent of the 50 people interviewed, whereas a fabula would be generally true of all 50 of them. Last, a narrative text is an expression of the story in language signs (e.g. words or pictures).

This narrative-centric approach moves away from interpretations of resilience purely in terms of economic valuation and seeks to gain a more comprehensive understanding of the emotional and mental coping costs involved, such as the negative experience of witnessing environmental damage caused to the Valley, and the loss of precious ecological sites. One example is that of the sacred Bagmati River which has run through the Kathmandu Valley for centuries, and which was turned into an open sewer and garbage dump for nearby factories (Muzzini & Aparicio, 2013).

A fabula is, therefore, a series of logically and chronologically related events that are caused or experienced by characters. It comprises of four elements – events, actors, time, and location. This paper uses the lived experiences method to collect these main elements.

Using a lived experiences approach, we investigate the subjective narratives of the delayed project; in particular, the self-reported perceptions of the residents' resilience or ability to cope with adversity which engenders the self-perception of resilience, so as to identify the extent to which these two hypotheses hold true in the case of Kathmandu.

Further, the fabula is also better able to explicate the paradoxes in human behavior, whereby empirical evidence does not always equate to corresponding explanations. A fabula allows us to peer underneath the surface of apparently resilient behavior to understand how and why responses occurred as they did, and how such responses can inform climate change and environmental policies.

Coping costs and the paradox of social resilience

Social resilience, therefore, has two elements – first, an agent within a certain policy context who is presented with adversity; and second, the agent's reaction to this adverse event. He is resilient insofar as he makes a positive adjustment.

Positive adaptation, the second construct, is usually defined in terms of behaviorally manifested social competence, or success at meeting some specific tasks (Waters & Sroufe, 1983; Masten, Best, & Garmezy, 1990; Luthar & Zigler, 1991; Masten & Coatsworth, 1998). High social competence is not, however, the only or preferred index used to define successful adaptation in resilience research; sometimes, the mere absence of emotional or behavioral maladjustment is appropriate (Rutter, 1999; Luthar *et al.*, 2000).

In this paper, these concepts are placed within the policy arena, such that social resilience is located within the research of Carpenter & Brock (2008), Walker *et al.* (2006) and Zolli & Healy (2012:9), who define resilience as a community's 'ability to maintain their core purpose to integrity in the face of dramatically changed circumstances'.

Extant literature has shown that this levies high and multidimensional costs on human life. It affects the everyday socio-economic decisions of households, and the act of coping with water scarcity also adds to psychological and emotional stress (Table 1).

Such coping costs are often given an economic valuation and calculated using revealed preferences (Katuwal & Bohara, 2007) or using water treatment behavior as a proxy (Katuwal & Bohara, 2011; Cook *et al.*, 2015).

In a paper by Pattanayak *et al.* (2005), it was shown that coping costs essentially contribute to another type of financial burden as a result of the inability of public water infrastructure to address households' water needs. An additional cost is thus required to supplement domestic water demand, using methods such as collecting, pumping, treating, and purchasing. These costs which are absorbed by households impose coping costs that can amount to 1% of the monthly income of households and thus represent the real but hidden costs of poor infrastructure service.

However, what is often less discussed are intangible costs – such as emotional costs – which are difficult to measure with existing quantitative methods. Stevenson *et al.* (2012) developed a new scale of water insecurity in order to understand women's emotional well-being in Ethiopia. They found that difficulty in obtaining sufficient and clean water introduces psychosocial stressors into women's lives such as physical illness, a conflict between family members or neighbors, shame at failing to adhere to normative standards of cleanliness and conduct, and absence during communal events.

Wutich & Ragsdale (2008) found that emotional distress amongst people in Villa Israel, Bolivia, is a result of uncertainty related to the regularity and sufficiency of water supply, as well as tension from neighbors and vendors, rather than a simple lack of water supply. In particular, that fear is related to irregular and insufficient water supply, whereas anger is related to interpersonal interactions involving water. The expression of different emotions in response allows researchers to obtain a clearer picture of the sources of emotional distress, even though emotional distress itself is hard to measure. Yet, because these emotions are so culturally specific and subjective, they cannot be measured by standard quantitative methods such as those employed by Cook *et al.* (2015) and Katuwal & Bohara (2007, 2011).

As such, the cost associated with these types of emotional labor must feature in any notion of social resilience.

Table 1. Coping costs, emotional, and economic.

No.	Source	Adversity/Positive outcomes	Intervention/Methodology	Findings/Summary
1	Wutich & Ragsdale (2008)	• High cost of water (~20% of average adult's income)/no positive outcome • Social tensions from borrowing water • Four emotions caused by water insecurity – fear, worry, and anger and bothering	• Random samples, 72 household heads and interviews • Binary data on emotions to create a 4-point scale of water • Related emotional distress using the Guttman scaling and multiple linear regression model	• Access to water and gender are significantly associated with emotional distress • Economic impact: half of household heads interviewed (47%) lost income because of the time it took to collect water and affect their survival (cattle and domestic gardens for vegetable, etc.)
2	Cook *et al.* (2015)	Sharing water with neighbors or using public wells or taps sometimes led to conflict	• Sixty percent of the 387 households were interviewed • Sampled households in four different 'Sub-locations' in Tigania West	• The total economic costs due to unreliable/distance water supply equal to 4.3% of the households reported income (US$38 per month). • Also, the coping costs depend on the size of the households
3	Stevenson *et al.* (2012)	• Tension arising from deciding between household chores and fetching water • Conflict with spouse and neighbors over water • Feeling ashamed for appearing unclean to others and failing to fulfill expectations of hospitality • Suspicions of neighbors stealing water	• 2009–2010 two phases of research (Multiple Regression Models, Focus group discussions) • Phase I – qualitative and designed to identify locally relevant experiences of water insecurity • Phase II – a quantitative survey to test the relationship between women's reported water insecurity and the Falk Self-Reporting Questionnaire (SRQ-F), a measure of psychosocial distress	• Water collection is women's responsibility in rural South Gondar, Ethiopia. This has impacts on women's health. Easy and reliable access to water contributes to healthy pathways to psychological well-being. However, stress of time, costs, and physical illness due to poor water supply contribute not only to family conflict but also social conflict, and failure to take part in customary communal events

Together, these provide a three-pronged methodological premise that undergirds our investigation of resilience:

1. Adverse event: Insecure and insufficient water supply.
2. Response: Coping with economic and emotional costs.
3. Fabula of resilience: Understanding responses to adverse events and perceptions of costs.

Kathmandu: water scarcity in the Kathmandu Valley

The Kathmandu Valley in Nepal is one of the fastest-urbanizing areas in South Asia, with an average urban population growth rate of 6% since the 1970s (UNDESA, 2012). These urbanization and population pressures place an enormous burden on the already weak infrastructure facilities in the Valley. Access to piped water in urban areas has decreased over the years from 68% in 2003 to 58% in 2010 (Muzzini & Aparicio, 2013). Further, access to water does not equate to water quality and much of the water that passes through pipes remain unsafe to drink. While daily domestic demand for water is around 220 million liters, the KUKL is only able to supply up to 100 million liters a day (Muzzini & Aparicio, 2013). There continues to be considerable under-investment in municipal water supply in the Valley, and policies geared toward addressing such deficits have encountered multiple delays due to inefficiency and political instability (Gurung *et al.*, 2017).

The Melamchi Water Supply Project (MWSP) was envisioned as a way to import raw water into the Valley to accommodate its long-term water needs arising from the burgeoning urban population. The Melamchi River, situated just outside the Valley, was identified as a potential water source, which could transfer water to the Kathmandu Valley via a 27.5 km diversion tunnel (Domènech *et al.*, 2013).

Since then, the MWSP has endured numerous obstructions and delays, owing to factors ranging from the withdrawal of lenders, operational obstacles, to civil war. In 2008, the public company Kathmandu Upatyaka Khanepani Limited (KUKL) was appointed on a PPP model to lead the construction of the tunnel, which formally began in 2009. Since then, little substantive progress has been made with multiple delays caused by contractual or financial problems, miscommunications between contractors and subcontractors, and natural disasters – including a devastating earthquake in 2015 that claimed 8,790 lives (Udmale *et al.*, 2016). Deadlines for project completion have been extended numerous times, from the initial goal of 2007 being delayed to 2012, then to 2016, and most recently October 2017 (The Himalayan Times, 2017a, 2017b).

At the time of writing, while 70% of the population have access to a piped water connection, many households receive water for only a few hours a day. As a result, many have coped in various ways – such as collecting water from public taps, investing in tubewells, and purchasing water from vendors and neighbors (Pattanayak *et al.*, 2005). The remaining 30% of the city's population that reside outside the piped network also use private tubewells, ancient stone water spouts, and supplies from tankers (Bhattarai & Conway, 2010).

As a result of this adaptation to a lack of piped water, informal water markets have grown exponentially (Domènech *et al.*, 2013). In 2004, there were about 60–70 tankers. By 2014, there were 210 firms operating about 700 water tankers in total (Gurung *et al.*, 2017). At the time of writing, some estimate that more than half of the city's water needs are being fulfilled by small private water tankers and bottled water vendors (Dixit & Upadhya, 2005), while others propose more conservative estimates; 8% of the peak and 4.8% of the off-peak season demand (Dongol *et al.*, 2012).

The supply comes from surrounding rural areas. Because of rising demand, water prices have increased – the market is now estimated to be worth 8 million USD per year (Pandey *et al.*, n.d.). This informal vending market represents a large part of the coping costs borne by the residents due to the continued delay in the Melamchi project.

Despite these rising costs, which place a tremendous burden on the majority of Kathmandu residents who live below the poverty line, people continue to accept and absorb the rising coping costs, with demand for private vendors increasing steadily. On average, the KUKL receives 60 complaints per month out of the approximately 1.9 million residents that it serves, most frequently regarding an absence of or limited supply; leaks and pollution; or unequal distribution (KUKL, 2008).

While frustrations have been expressed, there are minimal demonstrative protests over the water supply, as compared to protests over other issues, such as amendments to the constitution (Al Jazeera,

2017); air pollution (The Himalayan Times, 2017a, 2017b); or the rebuilding of monasteries (Ojha, 2017). It is, therefore, evident that while civil disobedience is not a foreign concept to the residents of Kathmandu, chronic water shortage does not rank as a protest-worthy cause.

Methods

The lived experience relies on a phenomenological premise that 'knowledge has its origin in experience'; in other words, 'objective realities' are built upon *subjective* perceptions (Sadala & Adorno, 2002:283) and there is 'no single "lived experience"' (Thomas *et al.*, 2008:328). As such, rather than focusing on the causal explanations usually elicited in surveys or interviews, phenomenology focuses on descriptions of experience, which are thought to bring the researcher into 'more direct contact' with the phenomenon in question (van Manen, 1984:1), as they challenge pre-given assumptions and bring into focus the experiences of minorities.

This method has been used in water research to provide a nuanced reading into different facets of water. For example, in her exploration of water as a distributed right, Parmar (2008) illustrates a striking difference between mainstream and rural communities' interpretation of the 'right to water'.

In constructing a fabula, a 'lived experiences' approach is useful as it provides a basis for constructing the logic of events as experienced by actors, within a certain time and location. By understanding the premise of their actions, it also provides an understanding of apparently irrational behavior – such as an over-estimation of one's ability to cope. This is central to the question in this paper.

However, a phenomenological approach has its limitations. First, unlike quantitative research methods, phenomenological studies often yield results that are non-replicable, due to the primacy given to individual experiences. A common criticism is that conclusions are, therefore, merely a subjective matter (Hycner, 1985), with little influence on policy scenarios where it is imperative to balance the needs of multiple stakeholders fairly. There is no assurance that the fabula constructed here is representative of social resilience as a whole.

In this paper, we focus specifically on an aspect of resilience that has been often elided; that of the emotional and cognitive costs of resilient responses. In doing so, we adopt a structural approach in seeking to understand how the actors, events, timing, and location of particular events coalesce in the manifestation of the paradox. We hope to express a narrative that explains the situation in Kathmandu from another angle and shed light on other situations where the paradox of resilience occurs.

The interviews were conducted in Kathmandu from May to July 2016. A total of 55 respondents participated in the exercise. The respondents are residents of Kathmandu and have an average age of 49 years. The youngest and oldest respondents were 18 and 83 years old, respectively. The respondents consisted of students, housewives, professionals, government servants, business owners, and retirees.

Two rounds of interviews were conducted for this study. The first, conducted on 26th May 2016, were short interview sessions (15–30 minutes) with the locals of Kathmandu to obtain a general understanding of their perception of the water supply situation in the city.

Based on the first round of interviews, a detailed set of questions were prepared and a subsequent interview session was carried out in the month of July 2016. A total of 36 local residents of Kathmandu were interviewed. Each session lasted an average of 60 minutes. Respondents had an average age of 41 years and ranged between 26 and 75 years old. Thirty-six percent of the interviewees were female.

All the interviews (amounting to 40 hours of tape) were transcribed and translated. The interviews were subsequently coded using the NVivo software.

Coding using NVivo involved identifying statements from each of the 36 interviews that expressed an idea, opinion, or fact in relation to five main themes, which emerged from the interviews. These themes form the 'units of relevant meaning' (Hycner, 1985:284) or significant and distinct topics (Sadala & Adorno, 2002) of the lived experience.

Our method follows Lindseth & Norberg's (2004) phenomenological hermeneutical approach. First, interview statements were read several times to get as close to the original meaning as possible. A 'naïve understanding' of participants' perceptions of water security was formulated based on our reading, allowing us to generate the five thematic statements in Table 2. The relevant statements were then categorized according to the different themes.

The statements were condensed into concise phrases based on their essential meaning, using the participants' own words as far as possible. These condensed statements were further sorted based on similarities and differences, generating sub-themes which formed a more complete picture of each theme.

Results and discussion

There are three main themes in the narratives, as shown in Table 3. These three themes correspond to our hypotheses (H1 and H2) in different ways.

First, there is a high cost to the current situation, a cost both in economic terms (Unit 1) and in opportunity costs, and behavioral changes (Unit 2). Participants are often acutely aware of the monetary cost of obtaining bottled and tanked water when piped water supply becomes intermittent, citing figures and comparing costs with household income. They also describe changes to daily routines to obtain water and heated arguments at public taps. Yet, despite the high costs, participants have adapted, accepting these costs as part of their lives. Paradoxically, the measures adopted further engender reliance on, and entrenchment of, the informal water market. This attitude of resignation and acceptance corresponds to our hypothesis of emotional numbing and an increased vulnerability despite an appearance of social resilience (H1).

On the other hand, the second and third themes represent uneasiness with the status quo and a rejection of both the informal market and existing government measures. Central to the second theme is the feeling that the current situation is not a stable equilibrium, requiring more drastic measures in the future (Unit 4). This theme has the largest number of statements and can be further broken up into two sub-themes: (i) that the current consumption pattern is unsustainable and (ii) that the problem has (and will) worsen over time.

Table 2. Result of the NVivo software analysis: number of statements per factor.

Units of meaning	Themes	No. of statements
1	We are paying a price high economic cost for poor water	29
2	Coping with water shortages is a part of our daily life	29
3	The government does not have the capacity to provide safe water	17
4	The situation will get worse over time	95
5	The water utility company in KMD is to be blamed	58

Table 3. Phenomenological hermeneutical analysis of KMD interviews: modified from.

Meaning unit	Condensation	Sub-theme	Main theme
I spend about Rs. 30,000 (S$390)/- per month for management of tanker, jar, and bottled water which I think lots of amount for water management. I pay Rs. 150 (S$2)/- per month for piped water and Rs. 3,000 (S$40)/- for tanker water. I feel I am paying a lot for tanker water	Recurring costs of compensating for poor water supply	Expensive alternatives to piped water	Economic cost of poor water supply
They are using the tap water by boiling and filter. Water becomes even more costly	Water has additional costs	Access to water	
Paying Rs. 5,500 (S$72)/- is a difficult thing for me as I have not that good income	It is hard to afford good water	Access to water	
There is no water supply for a long time they have to buy or collect water from public tap which needs more time and money	There is no secure supply of water	Water insecurity	
Price has changed hugely, for example, a tanker water used to cost 1,000 rupees in the past and it costs 1,500 rupees today, a bottle of water used to cost 15 rupees in the past, and it costs 20 rupees now, a jar used to cost 35 rupees and costs 50 rupees today	The problem has worsened over time	Unsustainability	
I pay Rs. 100 (S$1.3)/- bill for piped water, but I have to use the machine to pull water which increases lots of electricity bill at my home	Use of expensive electricity	Expense for water distribution	
Though my home is connected with piped water, there is no piped water supply at my home since a month, and I am fulfilling my water needs from a public tap which is called 'char dhara' situated about 300 m distance from my house	No water from pipes	Masked undersupply	Coping with water shortages is a part of daily life
I have seen people quarreling while standing in queues in public taps	Tensions observed from shortages	Welfare loss suffered	
Yes, this is not a recent problem, it is happening till around 2 decades	Long-standing problem	Adaptation to status quo	

People have no faith in political leaders, they have been hearing about the Melamchi project for two decades, but it still not completed	Lack of trust in the capacity of leaders	Long delay experienced	Lack of trust in government capacity
The municipal government only collects water bills from us, but they are not concerned about the water condition in our house	Money collected but no service rendered	Sense of unfairness	
Reasons for not carry out development activities is not lack of money but incapable staffs, politicizations, the intention of corruption, and recruitment of staffs through political parties	Poor view of public officials	Corruption	
There is no proper maintenance of the water pipeline connection	Poor operational maintenance	Poor performance	
Political leaders waste much money in dirty politics and non-profitable works, they organize unnecessary seminars regarding water supply but does nothing	Distrust in observed behavior	Corruption	
I am very angry with the KUKL and municipality for not caring. One can see old and unplanned water pipes in many places and water leakages in many places, but municipality and KUKL are silent on this	Anger against observed poor infrastructure	Poor performance	
People from all over Nepal are migrating into Kathmandu, current water resources are diminishing and no long-term water supply projects are in progress except the Melamchi project	Rising population	Stresses on water supply	Current consumption is unsustainable
If the water from the Melamchi project arrives, it may partially solve the water problem in Kathmandu but as the rapid growth in population and water demand and necessity, it will not be a permanent solution to the water problem in Kathmandu	Rising population	Even the Melamchi project will not be enough	

(*Continued.*)

Table 3. (*Continued.*)

Meaning unit	Condensation	Sub-theme	Main theme
Water used to come twice in a week (Tuesday and Wednesday) in my house before 2 years but water routine has been uncertain since 2 years (sometimes water comes and sometime not). In addition, water scarcity has risen after the devastating earthquake	Uncertain water supply and scarcity after the earthquake	Environmental hazards worsen water scarcity	The problem has worsened over time
Water price is going up and water availability is going down	Increasing water price	Increased water scarcity	
Because of growth in density of population, water problems are increasing since 8/9 years	Increased water problems due to population density	Population problems worsen water scarcity	
Before 8/9 years, there was not such a water problem in my house, piped water used to come every day	No water problems in past	Past water supply stable	
There was easy access to water 20 years ago, there used to have 24/7 water in the pipe, in well, and in public spouts or taps	Clean water was widely accessible	Past water supply stable	
Thirty years ago, pipelines, spouts, and natural wells were the sources where we could get clean water but this is not the case these days	Many sources of clean water in past	Past water supply stable	
I used to get water every 3 days before 5 years, after then I used to get in every week and nowadays, I am getting water every 10 days	Declining water access	Increased water scarcity	
Two years ago, piped water used to come once every 2 days, but now it comes only once a week today	Declining water access	Increased water scarcity	
KUKL is not serious about addressing the public water problem – staff there are more interested in their personal interests than the public good	Distrust of water utility company	Corruption	The water utility company in KMD is to be blamed
Being unable to do development activities are not due to lack of money but due to the unstable government, politicization in projects, politicization and corruption in KUKL	Money available but corruption hinders development	Corruption	
It collects money. But cannot invest	Money collected but no investment	Mismanagement of funds	

However, those areas where KUKL supplies water have been facing more water scarcity than our area	Larger problems in areas served by water utility company	Water utility company is the root of the problem	The water utility company in KMD is to be blamed
Maximum of unskilled staffs in KUKL are recruited by political parties, these are reasons for not being able to carry out development activities in by KUKL	Poor view of the water utility company	Corruption	
Staffs there are affiliated to political parties and they intend to corruption and achieve position than quality water supply	Distrust of water utility company	Corruption	
I feel angry about the current situation because KUKL supplies water one time in 10 days, which is very poor water supply	Anger against poor water supply	Water mismanagement	
KUKL collects enough money from water bills per month and gets supports from government and non-government organizations, so, the organization have sufficient money for development activities	Sufficient funding for development activities	Mismanagement of funds	
I think the KUKL should be abolished first. It would be better then	Abolishing the water utility company will improve the situation	Water utility company is the root of the problem	
KUKL staffs are more focused on commission and position than quality water distribution	Distrust of water utility company	Corruption	
I don't see proper management of water supply in Kathmandu	Poor management of water supply	Water mismanagement	

Population growth is viewed as a ticking time-bomb, and there is pessimism about the long-term ability of the Melamchi project to provide for growing water demand.

The last theme relates to a perception of incompetence on the part of the government. Many participants refer to corruption within the municipal government and the KUKL – including the personal greed of government and KUKL staff and deliberate mismanagement of funding as key reasons for failures in water provision. One statement even recommends the abolition of the KUKL as a start to dealing with the water problem. Above all, participants seem most emotional when describing government incompetence, emphasizing their anger at the perceived injustice or bitter resignation at the lack of progress on the issue of water supply.

The dissatisfaction with the status quo in the second and third themes accords with our second hypothesis of a 'rational calculus' of a weighing-up of relative costs and benefits of moving (H2). Participants recognize that the status quo of increasing water demand and poor water provision by the government and KUKL is unsustainable and that over time, they may be unable to bear the economic, social, and opportunity costs.

Participants' acute recognition of the unsustainable nature of the current situation, and their rational calculus of the increasing costs of poor water provision by the government and informal markets should lead to the conclusion that this is an important policy failure that requires a large-scale substantive response. However, there is little evidence of this conclusion across the narrative strands. This leads us to suspect that there has been an over-estimation of resilience.

The fabula consequently constructed comprises of three elements:

1. Coping (Hero Hypothesis): Although there is some blame on the government for the delay, the experiences also contain elements of surprisingly positive perceptions about the ability to cope. This contrasts with economic and scientific research about the high economic costs incurred.
2. Lack of alternatives to the status quo, mistrust, and lack of confidence in the capacity of the government. These elements lead to a lack of an identifiable villain and account for a relatively low valence of the public response. In this, we also find an explanation for why the status quo has persisted over a long period of time.
3. Tension and need for change: There are discourse elements which show that there are resignation and acceptance of the status quo; at the same time, there are also elements which acknowledge that the status quo is unsustainable. Therefore, a third element of the fabula is between the need for mitigation and that for adaptation. That is to say, a polyphony of two concepts of resilience: one which stands on stasis and another for change.

The fabula can be illustrated in a causal loop diagram (Figure 1).

As illustrated in the diagram, there are two processes affecting Kathmandu residents' resilience to water insecurity. There is a negative (balancing) feedback loop arising from resignation and acceptance of the status quo according to H1, which we call the 'Hero Hypothesis' as the residents perceive themselves to be coping with a difficult situation. This, however, is clearly an over-estimation.

From a simple water security point of view, the current situation arises from the informal nature of the supply, with over two-thirds of transactions taking place in cash, without any written contract. At the same time, there are no legal or institutional mechanisms that regulate the price, quantity, or quality of water sold by these vendors (Moench, 2001). Clearly, individuals cannot cope with these problems.

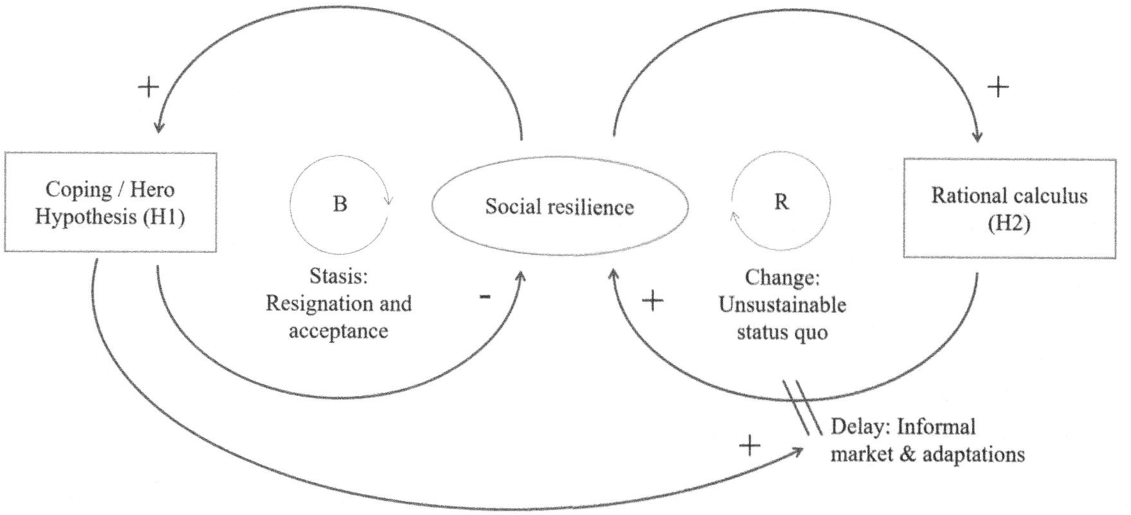

Fig. 1. Causal loop diagram of social resilience in Kathmandu.

As government-provided water accessibility and quality worsen, Kathmandu residents rely increasingly on informal water sources, producing an illusion of social resilience which in fact belies increasing vulnerability, as the unregulated informal water market becomes increasingly entrenched.

On the other hand, there is a positive (reinforcing) feedback loop arising from residents' dissatisfaction with the status quo. The rational calculus of increasing water costs should result in the recognition of policy failure and pressure for more drastic policy measures, eventually creating greater social resilience. However, the corrective effects of the rational calculus are delayed by the existence of the informal market and current adaptive measures, such as behavioral changes, as the presence of these stop-gap measures discourages comprehensive action on the issue.

From our analysis, it appears that both mechanisms of stasis and change associated with H1 and H2 can be integrated within an emerging line of research on resilience and its relationship with governmentality. This line of research highlights how subjects of resilience are 'responsibilized' (Kaufmann, 2013), as the concept of resilience implicitly accords normative values to personal or local community efforts to self-organize, learn from, and adapt to disruption while placing the responsibility of reacting to adverse circumstances on the subject.

As such, what we refer to as the 'paradox of resilience' in this paper also parallels an emerging body of research on the governmentality of resilience. Kaufmann's (2013) insight is instructive: the governmentality of the 'cultural village' concept creates community subjects who are incentivized to display resilience despite physical, economic, social, and emotional difficulties, to maintain the normative identity of the 'Cultural village' and the political and socio-economic benefits associated with it.

The governmentality of resilience is not fixed. As Kaufmann (2013) notes, subjects may display unexpected agency and act 'off-script'. In the case of Kathmandu, as population and environmental pressures on water supply increase over time, the rational calculus of economic, social, and opportunity costs of continuing to rely on the informal market and inconsistent government sources may become the dominant narrative over time, and disrupt the current notion of social resilience as 'coping' or individualized responsibility.

If that happens, the fabula and the casual diagram show us that the equilibrium will be disrupted severely.

Conclusions

In the end, what does a fabula of resilience tell us? A fabula is a construct, a structure of a story that can embed itself within different contexts. In constructing this fabula, we have elucidated three points about the concept of resilience.

First, we have shown resilience to be comprised of two important components. The first relates to a more static or passive conception of resilience, or an ability to withstand and work within existing institutions or circumstances. In our case, this is evidenced by way of the 'Hero' narrative, in which individuals inadvertently shoulder the coping costs associated with the failure of the Melamchi project, accepting these costs as an inevitable part of their lives.

The second is understood as a more active, reactionary conception of resilience, which indicates a willingness to depart from the status quo and challenge pre-existing institutions. This is indicated in the second and third elements of the fabula we have constructed, which acknowledges that the current situation is untenable and the desirability of breaking free from the status quo.

The tension between these two conceptions of resilience thus provides the impetus for a change. In a detailed examination of the discourses, one can perhaps appreciate how this is the only way for a community to remain viable in the face of adversity – to resist harmful changes while adapting to new and adverse environments.

The fabula also reveals the implications of the true costs of continued government inaction in what appears to be an emotionally muted situation. The temptation is for governments in such situations not to take action as there appears to be little political motivation or benefit to doing so. However, the lived experiences approach has shown that the present calm is engendered by an over-estimation of resilience and the perceived ability to cope. There is little guarantee that future crisis situations, such as political conflict or natural disasters, will not destabilize such a tenuous equilibrium. While the valiant efforts of citizens in coping with such adverse situations is not to be underestimated, the fabula reveals that there are real and significant pockets of vulnerability that exacts not only emotional costs but also psychological and emotional stresses that threaten the residents' future opportunities. Gong & Wu (2012) suggest that corruption in some developing countries is highly associated with low salaries. One policy to be considered could be to raise salaries to reduce widespread corruption.

A third contribution is that the fabula, by providing a relational map between the different elements, allows us to understand the processes that underlie the effects of both adverse environments, vulnerability, and protective factors. This is a point for future research.

As we have noted earlier, the identification of forces that show significant links with adjustment outcomes constitutes only the initial step for researchers in this field. The ultimate goal is to illuminate which potential mechanisms are implicated in the effects of these vulnerability or protective factors, such that appropriate directions for intervention can be derived.

The lived experiences method has revealed how the self-perceptions of resilience is fed by a rational calculus, which is due partly to the slow and incremental nature of the problem, and a perceived ability to pay coping costs (Hero Hypothesis). The lack of confidence in government capacity has meant that while the Hero can be identified, there is, at the moment, not a clear villain. Future study can be directed at how such heroes and villains are constructed within public narratives.

In conclusion, while phenomenological methods may require thoughtful implementation, there is significant potential for phenomenology to illuminate various dimensions of water as experienced by individuals, households, and human society as a whole. The potential for phenomenological approaches to change the researcher–participant relationship is also a significant draw, particularly when working with vulnerable individuals.

References

Al Jazeera (2017). *Nepalese opposition stages mass rally in Kathmandu, 7 Jan 2017*. Available at: http://www.aljazeera.com/news/2017/01/nepalese-opposition-stages-mass-rally-kathmandu-170106134544275.html.

Bal, M. (2009). *Narratology: Introduction to the Theory of Narrative*. University of Toronto Press, Toronto.

Bhattarai, K. & Conway, D. (2010). Urban vulnerabilities in the Kathmandu Valley, Nepal: visualizations of human/hazard interactions. *Journal of Geographic Information System 2*(2), 63.

Boissière, M., Locatelli, B., Sheil, D., Padmanaba, M. & Sadjudin, E. (2013). Local perceptions of climate variability and change in tropical forests of Papua, Indonesia. *Ecology and Society 18*, 4.

Carpenter, S. & Brock, W. (2008). Adaptive capacity and traps. *Ecology and Society 13*, 2.

Cook, J., Kimuyu, P. & Whittington, D. (2015). The costs of coping with poor water supply in rural Kenya. *Water Resources Research 52*(2), 841–859.

Dixit, A. & Upadhya, M. (2005). *Augmenting Groundwater in Kathmandu valley: Challenges and Possibilities*. Nepal Water Conservation Foundation, Kathmandu.

Domènech, L., March, H. & Saurí, D. (2013). Contesting large-scale water supply projects at both ends of the pipe in Kathmandu and Melamchi Valleys, Nepal. *Geoforum 47*, 22–31.

Dongol, R., Kansakar, L. K., Bajimaya, S., Maharjan, S. & Shrestha, D. (2012). Overview of water markets in the Kathmandu Valley. *Kathmandu Valley Groundwater Outlook, Kathmandu, Nepal* 100–111.

Eckersley, R. (2008). Nihilism, fundamentalism, or activism: three responses to fears of the apocalypse. *The Futurist 42*(1), 35.

Figueiredo, E., Valente, S., Coelho, C. & Pinho, L. (2009). Coping with risk: analysis on the importance of integrating social perceptions on flood risk into management mechanisms – the case of the municipality of Águeda, Portugal. *Journal of Risk Research 12*(5), 581–602.

Futerra, S. C. L. (2005). *The Rules of the Game: The Principles of Climate Change Communication*. Department for Environment, Food and Rural Affairs, London, UK.

Gong, T. & Wu, A. M. (2012). Does increased civil service pay deter corruption? Evidence from China. *Review of Public Personnel Administration 32*(2), 192–204.

González-Riancho, P., Gerkensmeier, B., Ratter, B. M., González, M. & Medina, R. (2015). Storm surge risk perception and resilience: a pilot study in the German North Sea coast. *Ocean & Coastal Management 112*, 44–60.

Guragai, B., Takizawa, S., Hashimoto, T. & Oguma, K. (2017). Effects of inequality of supply hours on consumers' coping strategies and perceptions of intermittent water supply in Kathmandu Valley, Nepal. *Science of The Total Environment 599*, 431–441.

Gurung, Y., Zhao, J., Kumar, K. C. B., Wu, X., Suwal, B. & Whittington, D. (2017). The costs of delay in infrastructure investments: A comparison of 2001 and 2014 household water supply coping costs in the Kathmandu Valley, Nepal. *Water Resources Research 53*(8), 7078–7102.

Holling, C. S. (1973). Resilience and stability of ecological systems. *Annual Review of Ecology and Systematics 4*(1), 1–23.

Hycner, R. H. (1985). Some guidelines for the phenomenological analysis of interview data. *Human Studies 8*(3), 279–303.

Kathmandu Upatyaka Khanepani Limited (KUKL) (2008). *Annual Report on Condition and Operation of the Service System*. 1 Falgun 2064–31 Asadh 2065.

Katuwal, H. & Bohara, A. K. (2007). Coping with unreliable water supplies and willingness to pay for improved water supplies in Kathmandu, Nepal. *Working Paper, University of New Mexico*.

Katuwal, H. & Bohara, A. K. (2011). Coping with poor water supplies: empirical evidence from Kathmandu, Nepal. *Journal of Water and Health 9*(1), 143–158.

Kaufmann, M. (2013). Emergent self-organisation in emergencies: resilience rationales in interconnected societies. *Resilience 1*(1), 53–68.

Laerhoven, F. (2014). When is participatory local environmental governance likely to emerge? A study of collective action in participatory municipal environmental councils in Brazil. *Environmental Policy and Governance 24*(2), 77–93.

Lindseth, A. & Norberg, A. (2004). A phenomenological hermeneutical method for researching lived experience. *Scandinavian Journal of Caring Sciences 18*(2), 145–153.

Luthar, S. S. & Zigler, E. (1991). Vulnerability and competence: a review of research on resilience in childhood. *American Journal of Orthopsychiatry 61*(1), 6.

Luthar, S. S., Cicchetti, D. & Becker, B. (2000). The construct of resilience: a critical evaluation and guidelines for future work. *Child Development 71*(3), 543–562.

Marshall, G. (2014). *Don't Even Think About It: Why Our Brains Are Wired to Ignore Climate Change*. Bloomsbury Publishing, New York and London.

Masten, A. S. & Coatsworth, J. D. (1998). The development of competence in favorable and unfavorable environments: lessons from research on successful children. *American Psychologist 53*(2), 205.

Masten, A. S., Best, K. M. & Garmezy, N. (1990). Resilience and development: contributions from the study of children who overcome adversity. *Development and Psychopathology 2*(4), 425–444.

Masten, A. S., Hubbard, J. J., Gest, S. D., Tellegen, A., Garmezy, N. & Ramirez, M. (1999). Competence in the context of adversity: pathways to resilience and maladaptation from childhood to late adolescence. *Development and Psychopathology 11*(1), 143–169.

Moench, Y. (2001). Water dynamics in a three part system: Investigation of the municipal, private and traditional public water supply systems in Kathmandu, Nepal. In: *Institute of Social and Environmental Transition and Nepal Water Conservation Foundation*, Kathmandu, Nepal.

Muzzini, E. & Aparicio, G. (2013). *Urban Growth and Spatial Transition in Nepal: An Initial Assessment*. World Bank Publications, Washington, DC.

Norgaard, K. M. (2011). *Living in Denial: Climate Change, Emotions, and Everyday Life*. MIT Press, Cambridge, MA.

Ojha, A. (2017). *Kathmandu Post: Monks take to streets to protest against delay. The Kathmandu Post, July 26*. Available at: http://bit.ly/2zMLOki.

O'Neill, S. & Nicholson-Cole, S. (2009). 'Fear won't do it' promoting positive engagement with climate change through visual and iconic representations. *Science Communication 30*(3), 355–379.

Pandey, V. P., Chapagain, S. K., Shrestha, D., Shrestha, S. & Kazama, F. n.d. Unpublished results. Groundwater markets for domestic water use in Kathmandu Valley: An analysis of its characteristics, impacts and regulations.

Parmar, P. (2008). Revisiting the human right to water. *Australian Feminist Law Journal 28*(1), 77–96.

Pattanayak, S. K., Yang, J. C., Whittington, D. & Bal Kumar, K. C. (2005). Coping with unreliable public water supplies: averting expenditures by households in Kathmandu, Nepal. *Water Resources Research 41*, 2.

Richler, J. (2017). Sociology: public inaction. *Nature Climate Change 7*(2), 96–96.

Rutter, M. (1999). Resilience concepts and findings: implications for family therapy. *Journal of Family Therapy 21*(2), 119–144.

Rutter, M. (2000). *Resilience Reconsidered: Conceptual Considerations, Empirical Findings, and Policy Implications*.

Sadala, M. L. A. & Adorno, R. D. C. F. (2002). Phenomenology as a method to investigate the experience lived: a perspective from Husserl and Merleau Ponty's thought. *Journal of Advanced Nursing 37*(3), 282–293.

Shaw, D., Scully, J. & Hart, T. (2014). The paradox of social resilience: how cognitive strategies and coping mechanisms attenuate and accentuate resilience. *Global Environmental Change 25*, 194–203.

Stevenson, E. G., Greene, L. E., Maes, K. C., Ambelu, A., Tesfaye, Y. A., Rheingans, R. & Hadley, C. (2012). Water insecurity in 3 dimensions: an anthropological perspective on water and women's psychosocial distress in Ethiopia. *Social Science & Medicine 75*(2), 392–400.

The Himalayan Times (2017a). *Protest against air pollution in Kathmandu, April 22 2017*. Available at: https://thehimalayantimes.com/multimedia/photo-gallery/protest-air-pollution-kathmandu/.

The Himalayan Times (2017b). *Water supply from Melamchi likely to be delayed by six months, August 4 2017*. Available at: https://thehimalayantimes.com/kathmandu/water-supply-from-melamchi-likely-to-be-delayed-by-six-months/.

Thomas, S. L., Hyde, J., Karunaratne, A., Herbert, D. & Komesaroff, P. A. (2008). Being 'fat' in today's world: a qualitative study of the lived experiences of people with obesity in Australia. *Health Expectations 11*(4), 321–330.

Udmale, P., Ishidaira, H., Thapa, B. R. & Shakya, N. M. (2016). The status of domestic water demand: supply deficit in the Kathmandu Valley, Nepal. *Water 8*(5), 196.

UNDESA (United Nations, Department of Economic and Social Affairs) (2012). *World Urbanization Prospects. The 2011 Revision*. UNDESA, New York.

Van Manen, M. (1984). *'Doing' phenomenological research and writing: An introduction (Monograph No. 7)*. University of Alberta, Edmonton, Canada.

Walker, B., Salt, D. & Reid, W. (2006). *Resilience Thinking: Sustaining People and Ecosystems in a Changing World*. Island Press, Washington, DC.

Waters, E. & Sroufe, L. A. (1983). Social competence as a developmental construct. *Developmental Review 3*(1), 79–97.

Wutich, A. & Ragsdale, K. (2008). Water insecurity and emotional distress: coping with supply, access, and seasonal variability of water in a Bolivian squatter settlement. *Social Science & Medicine 67*(12), 2116–2125.

Zolli, A. & Healy, A. M. (2012). *Resilience: Why Things Bounce Back*. Free Press, Simon & Shuster, New York.

Implications of the Melamchi water supply project for the Kathmandu Valley groundwater system

Bhesh Raj Thapa[a,b,*], Hiroshi Ishidaira[a], Maksym Gusyev[c], Vishnu Prasad Pandey[b], Parmeshwar Udmale[d], Masaki Hayashi[e] and Narendra Man Shakya[f]

[a]*Interdisciplinary Research Center for River Basin Environment (ICRE), University of Yamanashi, Takeda 4-3-11, Kofu, Yamanashi 400-8511, Japan*

**Corresponding author. E-mail: bthapa.ioe@gmail.com*

[b]*International Water Management Institute (IWMI) – Nepal Office, Durbar Tole, Pulchowk-3, Lalitpur, Nepal*

[c]*International Centre for Water Hazard and Risk Management (ICHARM), Public Works Research Institute (PWRI), Tsukuba, Japan*

[d]*Department of Civil and Earth Resources Engineering, Kyoto University, Kyoto 615-8540, Japan*

[e]*Department of Geoscience, University of Calgary, 2500 University Dr. NW Calgary, Alberta, T2N 1N4, Canada*

[f]*Department of Civil Engineering, Institute of Engineering, Tribhuvan University, Pulchowk, Lalitpur, Nepal*

Abstract

To meet the demand deficit in Kathmandu Valley, the Government of Nepal has planned to supply an additional 510 million liters per day (mld) of water by implementing the Melamchi Water Supply Project (MWSP) in the near future. In this study, we aim to assess the spatial distribution of groundwater availability and pumping under five scenarios for before and after the implementation of the MWSP using a numerical groundwater flow model. The data on water demand, supply infrastructure, changes in hydraulic head, groundwater pumping rates, and aquifer characteristics were analyzed. Results showed that groundwater pumping from individual wells ranges from 0.0018 to 2.8 mld and the average hydraulic head declined from 2.57 m below ground level (bgl) (0.23 m/year) to 21.58 m bgl (1.96 m/year). Model simulations showed that changes in average hydraulic head ranged from +2.83 m to +5.48 m at various stages of the MWSP implementation, and −2.97 m for increased pumping rates with no implementation of the MWSP. Regulation in pumping such as monetary instruments (groundwater pricing) on the use of groundwater along with appropriate metering and monitoring of pumping amounts depending on the availability of new and existing public water supply could be interventions in the near future.

Keywords: Groundwater management; Groundwater modeling; Kathmandu Valley; Melamchi

doi: 10.2166/wp.2019.084

Introduction

Kathmandu Valley's (KV) water resources management has been dependent on groundwater supply to meet increasing water demands due to the rapid population growth, deterioration of water-related infrastructure, improper land-use planning, and the rapid increase of built-up areas. Currently, the centralized potable water supply provided by the public agency, Kathmandu Upatyaka Khanepani Limited (KUKL), can only meet 19% (dry season) and 31% (wet season) of the valley's water demand, estimated at 370 million liters per day (mld) (KUKL, 2015) in 2015. The deficit between water demand and supply may increase up to 322 mld by 2021 (Udmale *et al.*, 2016). The deficit is currently filled by other sources such as rainwater harvesting, stone spouts, springs, and pumping from privately owned wells (Thapa *et al.*, 2016a, 2016b, 2018). Groundwater is considered to be a safe and reliable source of water, resulting in a steady increase in groundwater pumping from the valley's aquifers (Pandey *et al.*, 2010). For example, the KUKL (previously Nepal Water Supply Corporation) increased groundwater pumping from 2.3 mld in 1979 to 29.2 mld in 1999 (Pandey *et al.*, 2012) and the total estimated pumping including private withdrawals was 59.1 mld in 1998 (Metcalf & Eddy, 2000). In 2009, the estimated groundwater withdrawals reached 70.9 mld (Dhakal, 2010), resulting in a rapid decline of groundwater levels across the valley from 2.57 m to 21.58 m between 2003 and 2014 (KVWSMB, 2012; Shrestha, 2012; Gautam & Prajapati, 2014) below ground surface. Rapid urbanization may have exacerbated the decline by increasing impermeable surfaces in the valley's groundwater recharge areas (Cresswell *et al.*, 2001) and have increased the vulnerability of the aquifer system to pollution due to inadequate treatment of wastewater from urban areas (Pandey *et al.*, 2010; Shrestha *et al.*, 2016; Gautam *et al.*, 2017). Therefore, the current practices of groundwater use are unsustainable and could lead to irreversible groundwater quality deterioration and land subsidence in the valley (Shrestha *et al.*, 2017). Several studies have reported nitrate and *Escherichia coli* contamination in shallow groundwater as well as a high level of arsenic and iron in deep groundwater, making the valley's groundwater unsuitable for drinking without pretreatment (Warner *et al.*, 2008; Chapagain *et al.*, 2010; Pathak *et al.*, 2011; Tanaka *et al.*, 2012; Tamrakar & Shakya, 2013; Shrestha *et al.*, 2015).

Currently, there is a regulatory provision for acquiring licenses by paying an annual fee of 20,000 Nepalese rupees (NRs) (about 200 US dollars) for abstracting groundwater for non-domestic purpose from the deep wells. Each licensed client must install a bulk water meter to keep a monthly record of groundwater use and to renew the license on an annual basis. The groundwater abstraction can be completely shut down in the case of an operation without a license. The Kathmandu Valley Water Supply Management Board (KVWSMB) has demarcated three districts to issue groundwater abstraction licenses with a defined withdrawal limit: 2–15 liter per second (lps) in the northern district (safe zone); 2–7 lps in the central district (semi-critical zone); and 2–3 lps in the southern district (critical zone) (KVWSMB, 2014). Despite regular monitoring by the KVWSMB, these regulations are not fully enforced, causing licenses to often not be renewed and thus creating groundwater use records and fees to become not easily accessible. In addition, the number of shallow domestic wells are unknown due to the current provision of exempting licenses for groundwater withdrawals from dug wells up to 50 m below ground level (bgl) and tube wells with a diameter up to 10 cm (KVWSMB, 2014). As a result, the current governance of the valley's aquifer system has failed to curb groundwater demands or even be able to quantify existing groundwater withdrawal amounts in the valley.

To achieve sustainable water resources management, the Government of Nepal has planned to increase the valley's water supply by implementing an inter-basin water transfer program known as the Melamchi Water Supply Project (MWSP). The MWSP focuses on developing a bulk distribution system through improvement of the distribution network located mostly in the central groundwater district (CGD) and will be operated by the KUKL to supply water to the valley's existing users as well as to future users in new service areas in the Kathmandu Valley. The MWSP implementation will provide an additional 510 mld of water supply from the Melamchi River as an off-the-valley source in two phases: 170 mld by September 2016 (delayed by two years due to unforeseen circumstances including the 2015 Gorkha earthquake, shortage of fuel, and other project management issues) and 340 mld by 2023 (personal communication with procurement officer of the MWSP). Despite this planned surface water supply, the existing groundwater use in the valley is unlikely to be completely abandoned after the successful completion of the MWSP. This is due to high investments in groundwater supply infrastructure and the high pricing of KUKL water (NRs. 42.37/m^3). Therefore, there is a need to investigate the impacts of various groundwater pumping scenarios on the valley's aquifer system under pre- and post-MWSP conditions to evaluate appropriate water resource management actions.

This study is the first attempt to utilize a numerical modeling approach to investigate implications of the MWSP. Numerical groundwater flow models can provide quantitative answers to various questions regarding water resource management including impacts of changes in land use and land cover on water availability and of increased pumping on surface water features (Haitjema *et al.*, 2010; Gusyev *et al.*, 2012, 2013a, 2013b). A steady-state finite-difference groundwater flow model of the Kathmandu Valley watershed was set up using Visual MODFLOW (VMOD) Flex (Schlumberger Water Services, 2014). MODFLOW is a commonly used tool for groundwater resource assessment (McDonald & Harbaugh, 1988; Gusyev *et al.*, 2013a, 2013b; Anderson *et al.*, 2015). The results of model simulation are analyzed using geographical information system tools to estimate the change of simulated hydraulic heads between the baseline and various scenarios to: (1) assess the spatial distribution of groundwater pumping; (2) evaluate the behavior of hydraulic heads under different scenarios for pre- and post-MWSP conditions; and (3) suggest practical policy interventions for the sustainable utilization of groundwater resources in the valley based on the aquifer response. Despite the limitations of this study, these results help identify data needs in the watershed to improve the groundwater model and can serve as a baseline for future studies and sustainable management of the valley's groundwater.

Study area

The Kathmandu Valley watershed is located between 27°32′13″ and 27°49′10″N latitudes and 85°11′31″ and 85°31′38″E longitudes, and is surrounded by mountains. Elevations range between 1,212 and 2,772 m above mean sea level (amsl) (Figure 1). The watershed has a bowl-shape area of 664 km^2 and is drained by the Bagmati River and its tributaries (Bishnumati, Hanumante, Dhobi, Godabari, Balkhu, Kodku, Nakhhu, and Manohara). The outlet of the Bagmati River is located at the southern end of the watershed. The boundary of productive aquifer systems is outlined by the three KVWSMB groundwater districts with the new KUKL water supply service area located mostly in the central district (Figure 1). The population in the valley has increased from 1.1 million in 1991 to 1.6 million in 2001 to 2.5 million in 2011 (CBS, 2003, 2014a), which increased the water demand

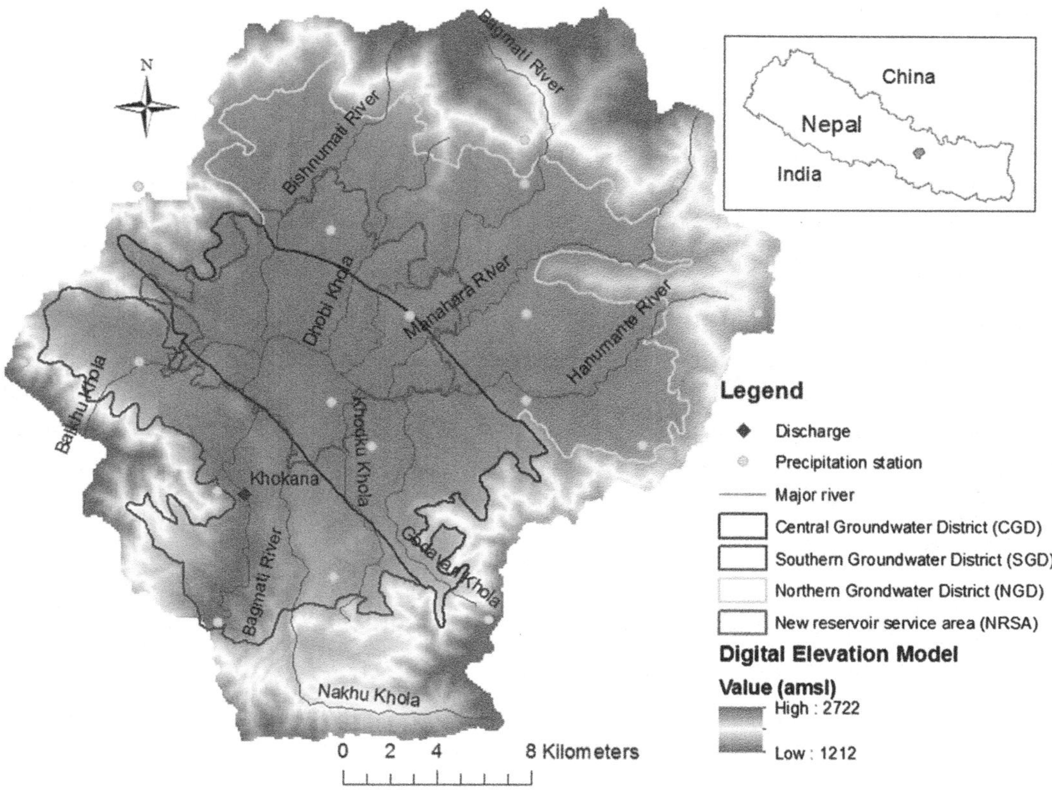

Fig. 1. The study area of the Kathmandu Valley watershed with precipitation and discharge gauging stations, groundwater observation wells, and the three groundwater districts and the boundary of the new reservoir service area for the Melamchi Water Supply Project.

from 155 to 370 mld (Thapa *et al.*, 2016a, 2016b, 2018). Agricultural lands in the valley have decreased from 62% to 42% between 1984 and 2000 (ICIMOD, 2007).

The KVWSMB groundwater districts are surrounded by low-permeability fractured rock formations and the outer periphery (hilly area) of the valley is covered by mixed forest. In the valley, the peri-urban area is partly agricultural and partly built up, while the central part of the valley is covered by built-up urban areas (Thapa *et al.*, 2016a, 2016b, 2018). The northern groundwater district (NGD) is situated on top of coarse unconsolidated deposits, which have high permeability and the highest potential for groundwater abstraction (Figure 1). The southern groundwater district (SGD) sits on top of a thick (200 m) confining unit of clay (JICA, 1990) with a deep confined aquifer composed of Pliocene sand and gravel with interbedded lignite, peat, and clay beneath the aquitard layer (Yoshida & Igarashi, 1984; Shrestha *et al.*, 2016). The CGD has a dual aquifer system separated by an aquitard of interbedded black clay and lignite up to 200 m thick (Yoshida & Igarashi, 1984). The shallow unconfined aquifer of the CGD is composed of up to 50 m thick Quaternary sand, with some discontinuities, interbedded silt and clay, and is underlain by thick (up to 200 m) black clay accompanied by some lignite and peat with low permeability (Shrestha, 2012). The deep confined aquifer is considered the main source of groundwater for the KUKL, private companies, and industries.

Materials and methods

Data collection

The data on population and its projection as well as on service area demarcation were extracted from published and unpublished reports by the KUKL and the MWSP. These data were used to estimate water deficit situations in the KUKL service area from estimated water demand and supply (Figure 2). The 2011 population census data were collected from the Central Bureau of Statistics (CBS, 2012). The record of population growth rates was obtained for each Village Development Committee (VDC) and Municipalities between 2001 and 2011 and projected rates from CBS for the year 2011 to 2031 (CBS, 2014b). Groundwater recharge rates were estimated for each hydrologic response unit, which is the smallest spatial unit having a unique combination of soil type, land cover and use, and slope within a sub-basin, using the physically based hydrological model Soil and Water Assessment Tool (SWAT) (Arnold *et al.*, 1998; Thapa *et al.*, 2017). The SWAT model was developed for the valley's watershed using precipitation, temperature, relative humidity, and solar radiation from year 2000 to

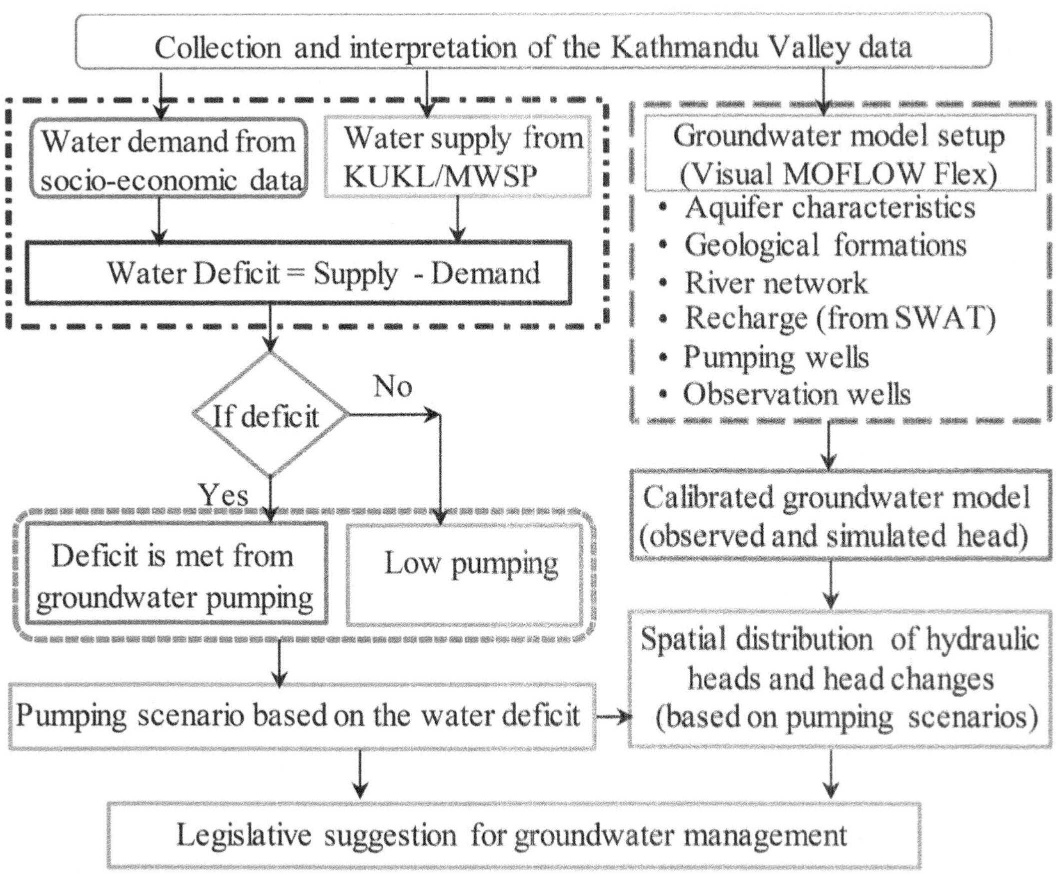

Fig. 2. Methodological framework of this study (KUKL is the Kathmandu Upatyka Khanepani Limited, MWSP is the Melamchi Water Supply Project).

year 2010 as input, and was calibrated using runoff data measured at Khokana gauging station (see location in Figure 1) (Thapa *et al.*, 2017). A surficial geology map was obtained from the Department of Mine and Geology (DMG/BGR/DOI, 1998). The latitude, longitude, and elevation for the three hydrostratigraphic units, also known as the shallow aquifer, the aquitard, and the deep aquifer, were obtained from Pandey & Kazama (2011). Information on pumping and observation wells was obtained from the Groundwater Resources Development Board (GWRDB).

The water deficit was estimated by Thapa *et al.* (2016a, 2016b, 2018) for the entire service area using a constant demand of 135 liters per capita per day (lpcd) and future demand for projected population was estimated using collected population data of each sub-area (at ward level) within the service area (CBS, 2012) with predicted rates (CBS, 2014a). The estimated water deficit was considered for the groundwater abstraction scenarios.

Pumping scenarios

Five different groundwater pumping scenarios were formulated based on the water deficit situation for the KUKL service areas considering three MWSP timelines as well as best and worst case scenarios. As per the estimated completion of the MWSP, water deficit situations were calculated for 2018 (after the completion of the first phase), 2023 (after the completion of the second phase), and 2030 (post-MWSP as well as the target year of achieving sustainable development goals (SDGs)). These scenarios generated for different time periods were implemented in a steady-state groundwater flow model, which was calibrated to groundwater heads and considered as a baseline in this study. Transient response of the valley groundwater system to a major change in pumping rates is an important factor, but it was not considered in the present analysis assuming an almost instantaneous equilibrium. The effects of pumping scenarios were investigated for the valley's reservoir service areas (RSAs) using the calibrated groundwater model with simulated hydraulic head distributions under the five scenarios. Simulation results were analyzed for each groundwater district to propose regulatory interventions for the sustainable management of groundwater resources in the valley. A brief narrative of the baseline and five scenarios is given below and summarized in Table 1.

Current rate of pumping (baseline). In 2016 (current situation), the KUKL provides water for the whole valley (old RSA) with rates ranging from 27 to 66 lpcd, which is generally less than the medium-term human requirement (WHO, 2005). The existing water deficit of domestic and industrial water use is matched by groundwater abstractions from privately owned wells. This scenario is the

Table 1. Groundwater pumping scenarios selected based on the potable water supply, demand, and deficits in the new and existing reservoirs' service areas.

Reservoir service area/Pumping scenario	Baseline scenario (current situation)	After 1st phase of MWSP, (S1)	After 2nd phase of MWSP, (S2)	Future (SDGs target), (S3)	Optimistic scenario, (S4)	Pessimistic scenario, (S5)
New Existing	Estimated pumping rate of 2016 (A)	50% of (A)	10% of (A) 125% of (A)	10% of (A) 150% of (A)	10% of (A)	150% of (A)

baseline case for pumping wells with current rates and these rates are used to generate other scenarios by using different percentages of pumping.

Pumping decreased by 50% of current rate (S1). The KUKL services more water than in 2016 after the first phase of the MWSP completed in 2018. The new RSA receives water at rates ranging from 70 to 130 lpcd, while the existing RSA receives 56 to 157 lpcd. In this scenario, we assume a 50% reduction in pumping for both new and existing RSAs as the future water demand is partly fulfilled.

Pumping at 10% of the baseline rate in the new RSA and 125% in the existing RSA (S2). The new RSA receives a sufficient amount of water after the second phase of MWSP is completed in 2023, but the situation in the existing RSA becomes worse than the baseline. Hence, this scenario considers pumping in the new RSA at 10% of the baseline rates and 125% of the baseline in the existing RSA. The 10% rate is considered 'minimal' pumping that may exist even after getting sufficient water from the KUKL utility. In the valley, it may be not possible to fully stop pumping because people may still rely on groundwater for some domestic purposes.

Pumping at 10% of the baseline rate in the new RSA and 150% in the existing RSA (S3). For the SDG's target year 2030, the new RSA has sufficient water supply with 10% pumping as in the previous scenario, but the situation in the existing RSA is much worse than in the year 2023. Hence, this scenario considers an increased pumping by 150% in the existing RSA.

Pumping at 10% of the baseline (or, optimistic scenario) (S4). This scenario considers the minimal pumping state in the entire valley. Assuming proper water redistribution in the KUKL service areas, all water demand is satisfied and there is no water deficit in the valley. Since the KUKL service areas already have physical infrastructure for groundwater pumping in place, there is a possibility of using groundwater for gardening and other non-domestic purpose due to poor water quality. This is considered to be an optimistic scenario and 10% of current pumping is assumed in both RSAs in this scenario.

Pumping 150% of baseline, (or, pessimistic scenario) (S5). This scenario considers a delay in the MWSP implementation and the water deficit is estimated to increase by 1.5 times of the current deficit. Hence, this is a pessimistic scenario and it considers an increased groundwater abstraction by 150% of the baseline scenario.

Model setup

A finite-difference groundwater flow model of the Kathmandu Valley watershed was set up in the VMOD Flex with a uniform grid size of 120 m (Figure 3). The VMOD model has three layers implemented with data obtained from Pandey & Kazama, (2011) to represent the shallow and deep aquifers separated by the aquitard. The spatially distributed geological formations were reclassified into six VMOD property zones in Layer 1 (one zone having the same hydraulic and storage property values) as shown in Figure 3(a) and implemented in each of three layers resulting in 18 zones, as shown in Figure 3(b) and 3(c). Note that zone 1 has inactive cells which represent a no-flow boundary of the model domain and is not considered in the model calculation (see the outside boundary of the modeled

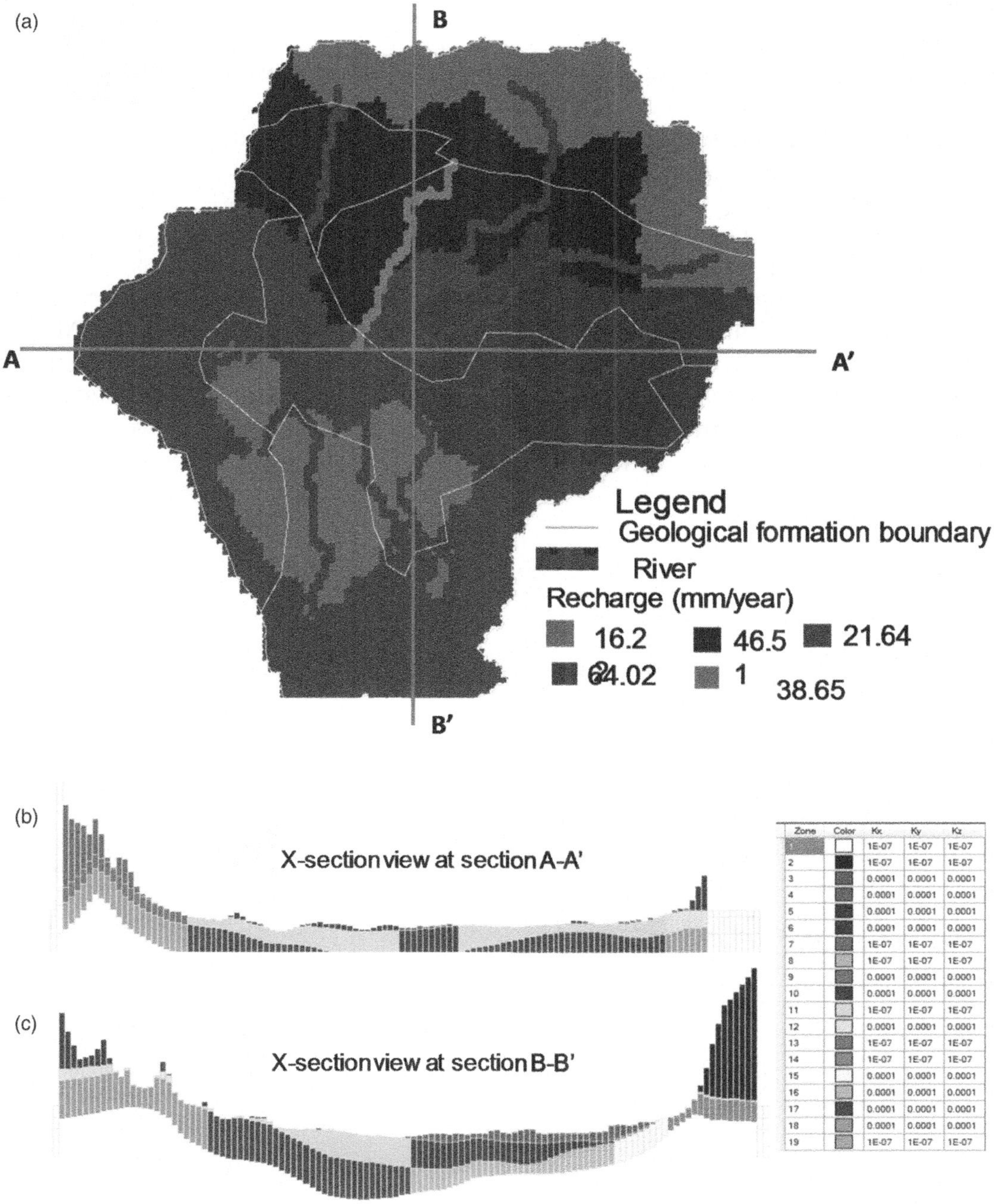

Zone	Color	Kx	Ky	Kz
1		1E-07	1E-07	1E-07
2		1E-07	1E-07	1E-07
3		0.0001	0.0001	0.0001
4		0.0001	0.0001	0.0001
5		0.0001	0.0001	0.0001
6		0.0001	0.0001	0.0001
7		1E-07	1E-07	1E-07
8		1E-07	1E-07	1E-07
9		0.0001	0.0001	0.0001
10		0.0001	0.0001	0.0001
11		1E-07	1E-07	1E-07
12		0.0001	0.0001	0.0001
13		1E-07	1E-07	1E-07
14		1E-07	1E-07	1E-07
15		0.0001	0.0001	0.0001
16		0.0001	0.0001	0.0001
17		0.0001	0.0001	0.0001
18		0.0001	0.0001	0.0001
19		1E-07	1E-07	1E-07

Fig. 3. Groundwater flow model setup in the Visual MODLOW Flex for Kathmandu Valley: (a) top view of aquifer recharge zones with river and drain networks and boundary of reclassified geologic formations utilized for property zones. Cross sections with hydraulic conductivity zone values are shown in (b) A–A′ and (c) B–B′.

valley's watershed and the cross sections in Figure 3). The VMOD zone budget was calculated for these 18 aquifer property zones in the VMOD model of the valley. Hydraulic conductivity values were assigned for each zone based on the available information on aquifer material properties and modified during the model calibration.

In Figure 3, five groundwater recharge zones and the surface water network were assigned in Layer 1 of the model. Nine perennial rivers were implemented with river package as surface water boundary conditions and one intermittent surface water feature was implemented as a drain. Shape files of the surface water network were imported and assigned as segments using river water and bed bottom elevations at the start and end of each segment, and river bed conductivity of 0.1 m d^{-1}. From the GWRDB, 19 groundwater observation wells with observed data from January 2003 were used for the model calibration and 379 pumping wells with available information (latitude, longitude, screen depth, and abstraction rates) were implemented in the VMOD model using the well package (Figure 4). The changes in water head over time were not considered, which is one limitation of this study. MODFLOW 2005 numerical engine was used and the rewetting option was active, allowing model cells to become dry or wet during the iterative solution of the steady-state flow equation. The model calibration was conducted manually by trial-and-error method to obtain an acceptable difference between observed and simulated heads. During the manual calibration, hydraulic conductivity values of one zone were

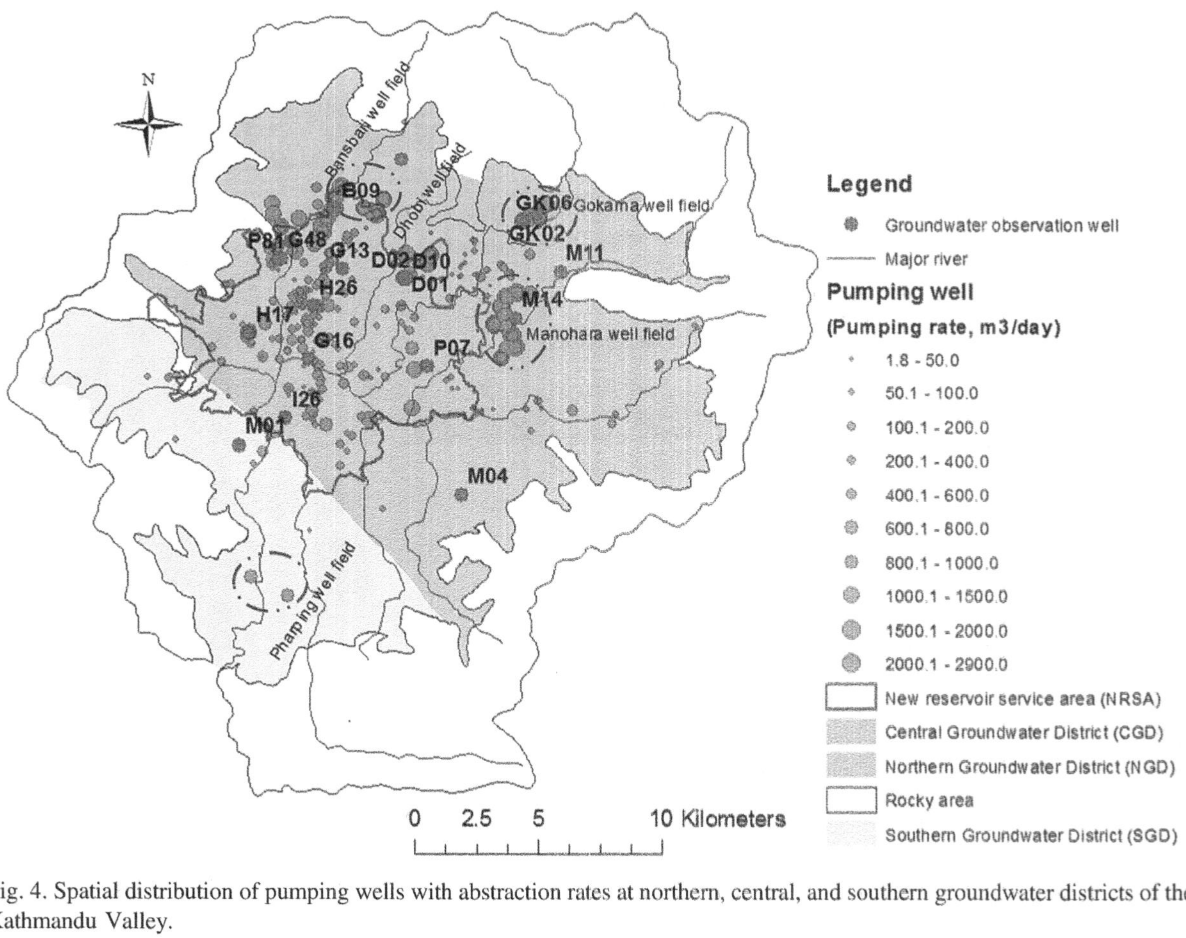

Fig. 4. Spatial distribution of pumping wells with abstraction rates at northern, central, and southern groundwater districts of the Kathmandu Valley.

adjusted to match observed heads of those wells in the zone. Water balance was evaluated using the VMOD zone budget to confirm realistic river water losses and gains of the model simulation. The same procedure was conducted iteratively for all 18 zones.

Results and discussion

Groundwater pumping

In 2012, the GWRDB (McDonald & Harbaugh, 1988) reported that most of the existing 759 deep tube wells had been abandoned, and only some were still in use by various users, such as hotels, embassies, government offices, and carpet and brick companies. It is a limitation of this research that only 379 pumping wells listed in the GWRDB inventory and their pumping rates in 2009 were considered while there could be many more new wells, some of which may have been already abandoned. The majority of the 379 pumping wells are located in the central part of the valley, where the KUKL is planning to supply water from off-the-valley source via the MWSP, with 154 wells in the NGD, 212 wells in the CGD, and 13 wells in the SGD. The estimated groundwater abstraction from the valley was 143 mld, consisting of 102, 38, and 3 mld from the central, northern, and southern groundwater districts, respectively. Abstraction rates from individual wells ranged from 0.00018 to 2.8 mld, with an average of 0.65 mld for the NGD, 0.00028–1.1 mld with an average of 0.18 mld for the CGD, and 0.00042–0.86 mld with an average of 0.24 mld for the SGD (Figure 4). Higher pumping rates were mainly from the KUKL production wells and three wellfields in Manohara, Bansbari, and Dhobi. The average rate of hydraulic head decline in the valley was 0.69 m/year for the 11 year (2003–2014) period with 1.12 m/year in the NGD, 0.72 m/year in the CGD, and 0.23 m/year in the SGD. For the NGD, the observed decline of hydraulic head is 12.35 m and is 21.58 m at well B13 and 3.12 m at well M12 in the 11 year (2003–2014) period. The groundwater head of CGD wells declined by 5.05 m at well H17, 6.33 m at well P81, 11.6 m at well I26, and 8.76 m at well M04, and the district average was 7.93 m. The SGD has only one observation well, which is M01 and has a reported 2.57 m decline of its groundwater head for the 11 year (2003–2014) period.

Model calibration with baseline scenario

Figure 5 shows the correlation between observed hydraulic heads in January 2003 and the heads generated by the steady-state simulation using 2009 pumping rates. The calibrated groundwater model had a correlation coefficient of 0.70 with observed data, which is a reasonable model performance considering the limitation of available data and the large uncertainty in subsurface information. Despite a reasonable overall model fit, the errors are not randomly distributed across the valley; the simulated hydraulic heads for the wells in low-elevation areas have large positive biases and those in high-elevation areas have negative biases (see locations of the observation wells in Figure 4). The minimum head difference of 3.7 m was observed at well M12 and maximum head difference of 21.47 m at well P07, which is screened in Layer 2, while another well G16 screened in Layer 2 shows a better fit. Two wells, D10 and D02, located in close proximity to each other in the NGD, demonstrate an underestimation of simulated heads in Layer 1. Observation wells H26 and G13, screened in Layer 3, showed an overestimation of groundwater heads and may be due to the oversimplification of the hydrogeology on the western side

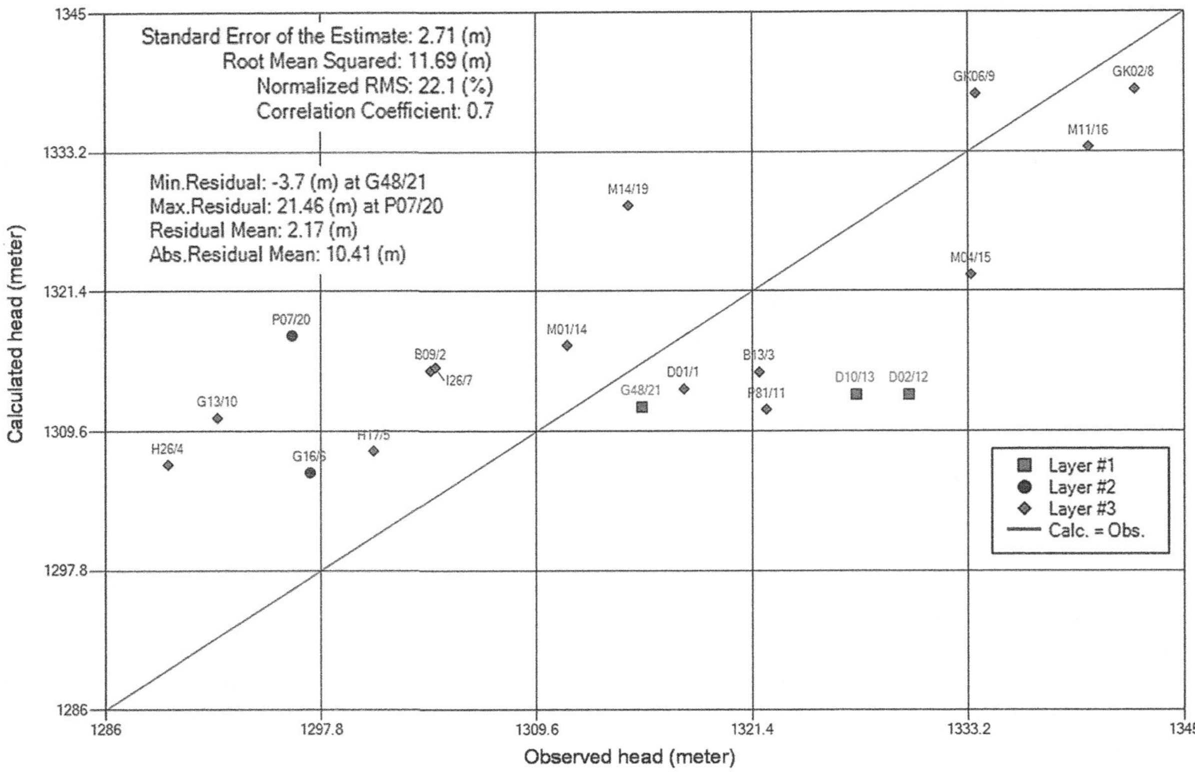

Fig. 5. Calibration of hydraulic heads for January 2003 using unpublished observed values from GWRDB (Groundwater Resources Development Board).

of the valley. These groups of observation wells with large simulated differences should be the focus of future studies to improve the model performance locally as well as across the valley. Figure 6 shows the spatial distribution of simulated hydraulic heads in the shallow (Layer 1) and deep (Layer 3) aquifers. A cone depression is observed in both Layers 1 and 3 in the CGD, but they are slightly offset due to the different rate of pumping in each layer. For scenario analysis, results are considered as satisfactory, but the groundwater flow model could be further improved with additional data in future studies.

Analysis of scenario simulations

Figure 7 shows changes in simulated hydraulic heads between the baseline and the five scenarios summarized in Table 1 in Layer 3 (Figure 7(a)). The results of the first scenario show an average increase of hydraulic heads by 3.37 m across the three groundwater districts and by 3.20 m, 2.77 m, and 4.49 m in NGD, CGD, and SGD, respectively (Figure 7(b)). The reduction of pumping is considered after the implementation of the first phase of the MWSP and indicates a greater increase in the northwestern part than in the central part of the valley.

In the second scenario (Figure 7(c)) after the implementation of the second phase of the MWSP, the new RSA receives a sufficient amount of water and hence there is a possibility to reduce pumping,

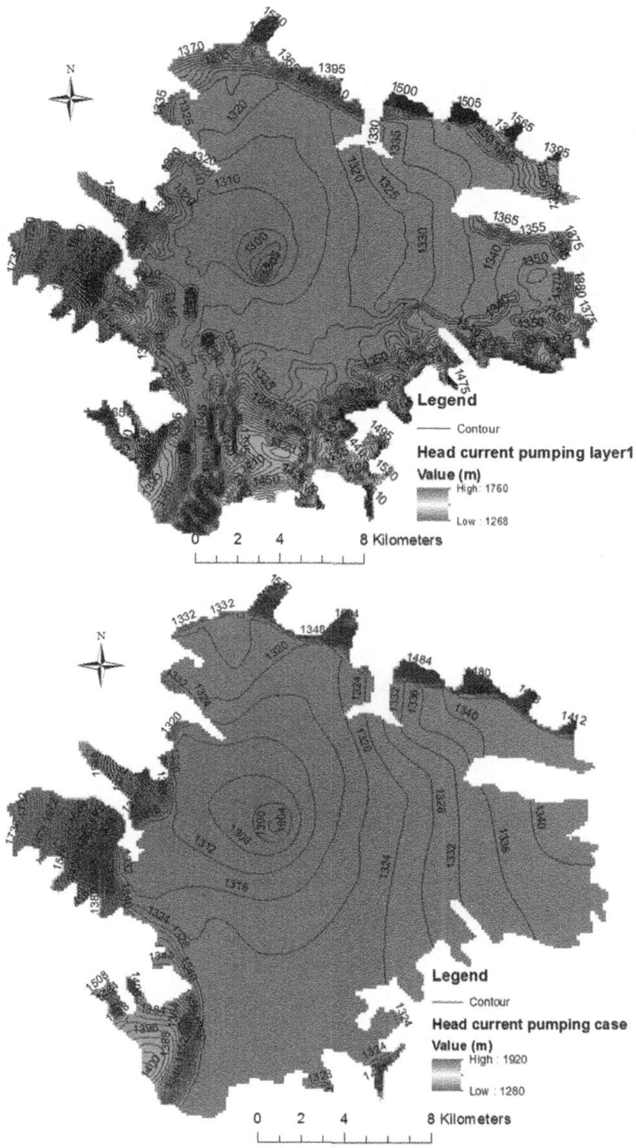

Fig. 6. Spatial distribution of simulated hydraulic heads in the shallow (top) and deep (bottom) aquifers in the three groundwater districts within the Kathmandu Valley using the calibrated model. Head values are in meters above mean sea level.

perhaps to the minimal pumping condition. In contrast, the water deficit becomes worse in the existing RSA, leading to an increase of pumping by 25%. As a result, there is an increase (up to 11 m) in hydraulic heads inside the new RSA, mainly in the central-western part of the valley. The increase of pumping by 25% in the existing RSA increases the head (up to 7 m) in the north-western part of the valley at nearby Bansbari well field, but it results in a decrease of heads in the north-eastern part of the valley (Figure 7(c)). The average increase of hydraulic heads is 2.24 m, 3.59 m, 4.25 m in the NGD, CGD, and SGD, respectively. The average increase of head is 3.15 m for the entire valley due to a major

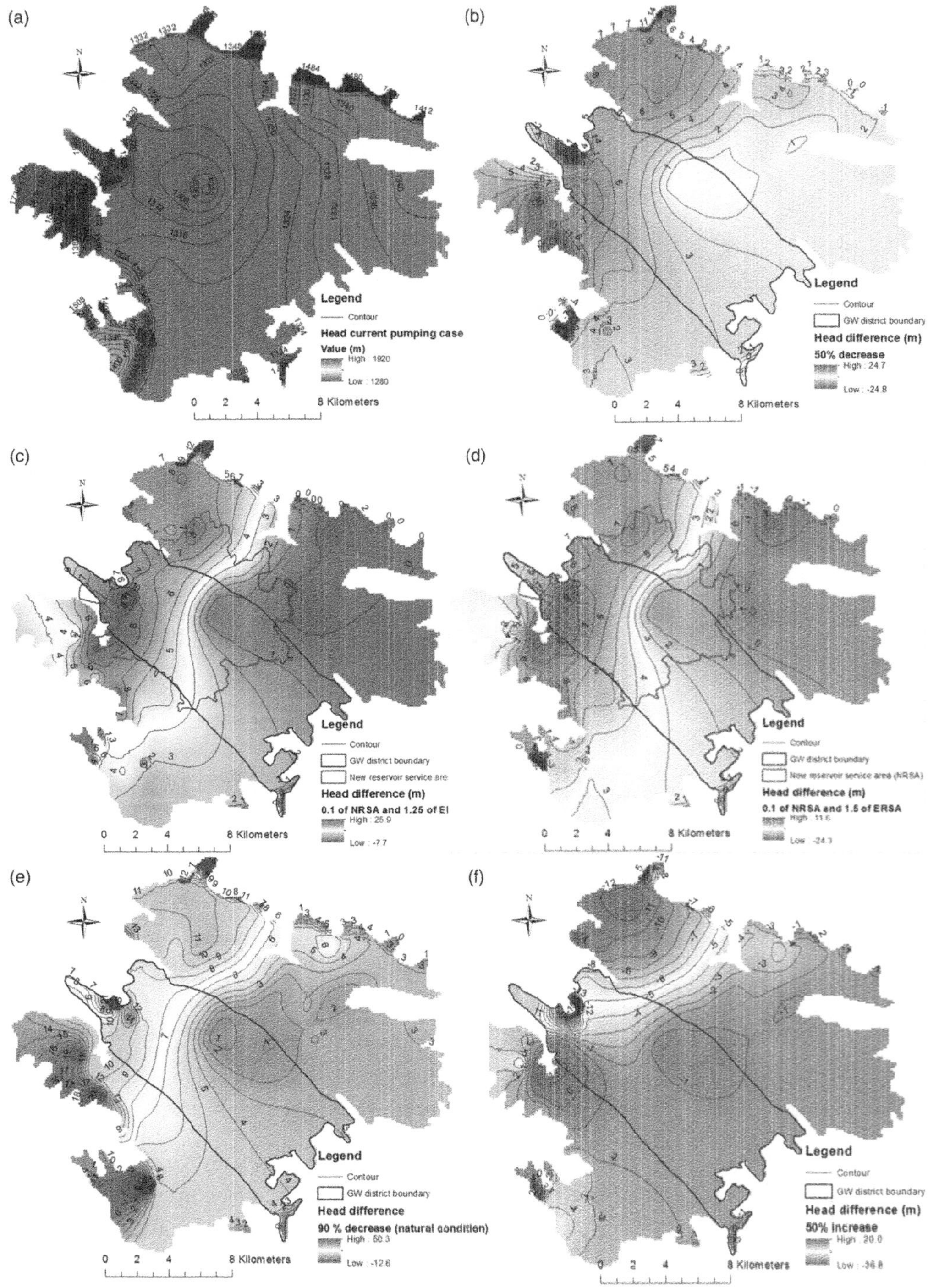

Fig. 7. Simulated heads in the baseline scenario (a) and head changes between the baseline and (b) 50% reduction, (c) minimal (10%) in the new RSA and 25% increase in the existing RSA, (d) minimal (10%) in the new RCA and 50% increase in the existing RSA, (e) optimistic, and (f) pessimistic scenarios.

decrease in pumping inside the new RSA, where most of the pumping wells occur, and the water deficit is substantially reduced inside the new RSA in this scenario.

In the third scenario, in the existing RSA, the situation would deteriorate due to the unmet total demand and increased pumping by 50% in the existing RSA. In this scenario, hydraulic heads decrease in the north-eastern part of the valley and increase in the central and north-western parts (Figure 7(d)). The average change of hydraulic heads is 1.62 m, 3.47 m, and 4.2 m in the NGD, CGD, and SGD, respectively. The average change of heads for the valley is 2.83 m. These results indicate that there is a substantial impact on the groundwater system after the completion of the MWSP, especially if proper regulatory mechanisms are able to curb pumping from the deep aquifer by substituting it with the piped water supply to meet any water deficit.

In the optimistic scenario, the MWSP is completed in the stipulated time and the water supply is redistributed across all sub-districts of both new and existing RSAs, fully satisfying all water demands. The minimal pumping in the entire valley leads to increases in hydraulic heads by 5.48 m, on average, and 5.06 m for the NGD, 4.83 m for the CGD, and 7.11 m in SGD (Figure 7(e)). The highest increase in SGD is due to a major reduction in pumping in two wells in the southern part (Pharping well).

For the pessimistic scenario, there will be a severe water deficit with a possible increase in pumping by 150% of the current rate in the entire valley if the MWSP is not completed on time. The average decrease of hydraulic heads is −2.97 m for the KV and is −4.0 m, −2.19 m, and −2.19 m in the NGD, CGD, and SGD, respectively (Figure 7(f)).

Limitation of the current study

Despite the availability of advanced modeling technology with available local data, the current model setup is only the first attempt to represent the complex aquifer system of the Kathmandu Valley. The first limitation of this study is the lack of updated groundwater abstraction information from active pumping wells because the pumping rate data were available only for 379 out of 759 existing wells and estimated from an old record from 2009. Using a more current and comprehensive data set would improve our estimates of the decline of hydraulic heads and identification of the areas of potential concern. A sparse observation well network and the lack of river discharge data were the second major limitations of this study. To calibrate the groundwater flow model, the 2003 hydraulic head data were used as the most complete data set, while river discharge data were unavailable to calibrate surface–groundwater interactions on the majority of perennial streams. Two river gauging stations allowed a match of river losses to groundwater between simulated and observed discharges on river reaches, but this is insufficient for the rest of the watershed, especially in groundwater recharge areas. In this study, we simplified the surface geological information into six subgroups assigned to all three depth layers. The mountainous area in the active model domain based on perennial rivers was assumed to be a low-permeability zone. Utilizing isotope tracers such as tritium in wells and river water during the base flow period will help to quantify groundwater contributions from the low permeable areas to the perennial streams and the aquifer system, respectively (Gusyev *et al.*, 2016). Having more detailed geological and hydrogeological information should resolve problematic areas in future models by using more vertical layers with finer spatial grids. In addition, we assumed a steady-state situation in the aquifer system which was necessary for evaluating change in groundwater heads in management scenarios. However, the decline of groundwater heads at different rates is seen in all available observation wells and a transient model will be useful for determining the temporal responses of the groundwater flow system for pumping in future studies.

Summary and recommendation

This study presents an investigation of water resources management scenarios in the Kathmandu Valley using a groundwater flow model developed with limited available data. Groundwater models are useful tools that can make use of available data and guide new data collection in critical areas for improving simulation results, while answering practical questions regarding water management. In model simulations representing various stages of implementation of the MWSP, estimates of groundwater abstraction rates played a critical role in the analysis of the impacts of the MWSP. The model was used to simulate the spatial distribution of hydraulic heads for the different scenarios and the results were used to evaluate the implication of changes in pumping rates and recommend policy options. The first phase of the MWSP will partially fulfill water demand of both new and existing RSAs, while after the second phase of the MWSP, the water demand of the new RSA is fully met but the existing RSA still faces a water deficit. Despite getting sufficient utility water, there is still a possibility to abstract groundwater using available infrastructure. The spatially averaged changes in hydraulic head ranged from +2.83 m to +5.48 m in various stages of the MWSP implementation and −2.97 m for increased pumping rates with no implementation of the MWSP.

These findings demonstrate a potential improvement in groundwater management with the implementation of the MWSP and proper regulation. There may be inequality in access of potable water, which could lead to conflicts among the groundwater users and regulators. In this situation, subsidizing water pricing based on the water deficit situation in the RSAs may be one solution. Detailed assessments of implementation of subsidizing water pricing needs to be considered in future studies. Based on these results, it is recommended that, after the first phase of the MWSP, the KVWSMB motivates groundwater users to voluntarily reduce their pumping rates by up to 50% for the sustainable management of groundwater. However, after the second phase of the MWSP, appropriate regulation on groundwater pumping needs to be based on the accessibility of piped water supply and enforced with appropriate regulatory and monetary mechanisms. Such regulatory mechanisms could be monetary incentives to install meters to monitor pumping rates. For example, the KVWSMB may aim to impose a higher price for groundwater pumping licenses in the new RSA, medium prices in the existing RSA, and nominal charges in the areas having neither new nor existing public water supply based on the price and availability of water. Although the installation of bulk meters is expensive for consumers and regular groundwater use monitoring is not an easy task for the KVWSMB, these data are required for policy formulation, zoning of groundwater potential, groundwater vulnerability mapping, and regulating groundwater withdrawals.

Currently, several rules, regulations, and strategies exist for regulating water supply and sanitation in the valley; however, the situation has not improved over the past decades. The historical trend of observed hydraulic heads demonstrates a steady decline of 0.69 m/year due to increasing groundwater abstraction in the valley. Combining the current depletion of the aquifer system with increasing water demand in the future will result in the overexploitation of groundwater resources. Hence, various policy instruments are necessary for the effective implementation of spatially differentiating pricing policy. The implications of various policy regulations to recommend differing pricing policy on the valley's groundwater system need to be identified in future studies with a coupled economy and groundwater model.

The MWSP is the key project to combat the present water deficit, but investigations into the impact of the MWSP on the groundwater system have been lacking. In this regard, the current study is the first attempt to evaluate the implications of the MWSP on the valley's groundwater system linking with the potable water deficit calculated from water demand and supply. Therefore, this study provides answers to water management questions, such as 'What may be the impacts of possible change in pumping in different stages of MWSP implementation?' In addition to the MWSP implementation, there will be a need to evaluate impacts of climate and land-use changes in the valley's groundwater system. These changes could affect the permeability of the urban landscape and increase wastewater discharge into rivers. Therefore, the spatial dynamics of these changes should be incorporated in future research.

Acknowledgments

The authors would like to acknowledge JST-JICA SATREPS Project in Nepal (PI: Futaba Kazama, University of Yamanashi) and Ministry of Education, Culture, Sports, Science and Technology, Japan (MEXT) for supporting this study financially. The authors also express their thanks to several organizations in Nepal: Department of Hydrology and Meteorology (DHM); Ground Water Resources Development Board (GWRDB); Melamchi Water Supply Project (MWSP); and Kathmandu Valley Water Supply Management Board (KVWSMB); Center for Research for Environment, Energy Water (CREEW) for sharing useful data and information. The authors are also grateful to Sanjeev Bikram Rana, Executive Director, KVWSMB; Ram Prasad Kharel, Procurement Officer, MWSP; Tilak Mohan Raj Bhandari, Technical Manager, KUKL; Surendra Raj Shrestha, Senior Divisional Hydrogeologist, GWRDB; Anoj Khanal, Hydrogeologist, KVWSMB; Dr Jun Magome; and Anil Aryal for their kind support during data collection and interpretation. Constructive comments by two anonymous reviewers improved the quality of the paper.

References

Anderson, M., Woessner, W. W. & Hunt, R. (2015). *Applied Groundwater Modeling, Simulation of Flow and Advective Transport*, 2nd edn. Elsevier, Amsterdam, The Netherlands.

Arnold, J. G., Srinivasan, R., Muttiah, R. S. & Williams, J. R. (1998). Large area hydrologic modeling and assessment part I: model development. *Journal of the American Water Resources Association 34*(1), 73–89.

CBS (2003). *Population Monograph of Nepal: Vol.-2*. Central Bureau of Statistics, Kathmandu, Nepal.

CBS (2012). *National Population and Housing Census 2011 (National Report)*. Government of Nepal, Central Bureau of Statistics, Kathmandu, Nepal.

CBS (2014a). *National Population and Housing Census 2011 (Population Projection 2011–2031)*, Vol. 8. Central Bureau of Statistics, Kathmandu, Nepal.

CBS (2014b). *National Population and Housing Census 2011 (Village Development Committee/Municipality)*, Vol. 6. Central Bureau of Statistics, Kathmandu, Nepal.

Chapagain, S. K., Pandey, V. P., Shrestha, S., Nakamura, T. & Kazama, F. (2010). Assessment of deep groundwater quality in Kathmandu Valley using multivariate statistical techniques. *Water, Air, and Soil Pollution 210*(1–4), 277–288.

Cresswell, R. G., Bauld, J., Jacobson, G., Khadka, M. S., Jha, M. G., Shrestha, M. P. & Regmi, S. (2001). A first estimate of ground water ages for the deep aquifer of the Kathmandu Basin, Nepal, using the radioisotope chlorine-36. *Ground Water 39*(3), 449–457.

Dhakal, H. P. (2010). *Groundwater Depletion in Kathmandu Valley. Presentation in Water Environment Partnership in Asia-Nepal Dialouge, Kathmandu), 14 December 2010.* Available from: http://www.wepa-db.net/pdf/1012nepal/09.pdf.

DMG/BGR/DOI (1998). *Hydrogeological Condition and Potential Barrier Sediments in the Kathmandu Valley.* Technical Cooperation Project Environmental Geology, Final Report.

Gautam, D. & Prajapati, R. N. (2014). Drawdown and dynamics of groundwater table in Kathmandu Valley, Nepal. *Open Hydrology Journal 8*, 17–26.

Gautam, D., Thapa, B. R. & Prajapati, R. N. (2017). Indigenous water management system in Nepal: cultural dimensions of water distribution, cascaded reuse and harvesting in Bhaktapur City. *Environment, Development and Sustainability 20*(4), 1889–1900.

Gusyev, M. A., Toews, M. W., Daughney, C. J., Hong, T., Minni, G., Fenemor, A., Ekanayake, J., Davie, T., Basher, L. & Thomas, J. (2012). Modelling groundwater abstraction scenarios using a groundwater-river interaction model of the Upper Motueka River catchment. *Journal of Hydrology New Zealand 51*(2), 85–110.

Gusyev, M. A., Haitjema, H. M., Carlson, C. P. & Gonzalez, M. A. (2013a). Use of nested flow models and interpolation techniques for science-based management of the Sheyenne National Grassland, North Dakota, USA. *Ground Water 51*(3), 414–420.

Gusyev, M. A., Toews, M., Morgenstern, U., Stewart, M., White, P., Daughney, C. & Hadfield, J. (2013b). Calibration of a transient transport model to tritium data in streams and simulation of groundwater ages in the western Lake Taupo catchment, New Zealand. *Hydrology and Earth System Sciences 17*(3), 1217–1227.

Gusyev, M. A., Moregenstern, U., Stewart, M. K., Yamazaki, Y., Nishihara, T., Kuribayashi, D., Sawano, H. & Iwami, Y. (2016). Application of tritium in precipitation and baseflow in Japan: a case study of groundwater transit times and storage in Hokkaido watersheds. *Hydrology and Earth System Sciences 20*(7), 3043–3058.

Haitjema, H. M., Feinstein, D. T., Hunt, R. J. & Gusyev, M. A. (2010). A hybrid finite-difference and analytic element groundwater model. *Ground Water 48*(4), 538–548.

ICIMOD, MoEST, & UNEP (2007). Kathmandu Valley Environment Outlook, Kathmandu, Nepal. In: *International Centre for Integrated Mountain Development (ICIMOD)*, Kathmandu, Nepal.

JICA (1990). *Groundwater Management Project in Kathmandu Valley.* Final report, main report and supporting reports.

KUKL (2015). *Kathmandu Upatyaka Khanepani Limited Annual Report.* Kathmandu Upatyka Khanepani Limited, Kathmandu, Nepal.

KVWSMB (2012). *Groundwater Resources Management Policy.* Kathmandu Valley Water Supply Management Board, Kathmandu, Nepal.

KVWSMB (2014). *Guidelines for Extraction, Utilization and Issuing License for Groundwater in Kathmandu Valley. 2014.*

McDonald, M. G. & Harbaugh, A. W. (1988). *A Modular Three-Dimensional Finite Difference Ground-Water Flow Model.* Techiques of Water-Resources Investigations, USGS, Denver, Book 6, p. 588.

Metcalf and Eddy (2000). *Urban Water Supply Reform in the Kathmandu Valley (ADB TA Number 2998-NEP).* Completion reports, vols I, II, Executive summary, main report and Annex 1 to 7.

Pandey, P. V. & Kazama, F. (2011). Hydrogeologic characteristics of groundwater aquifers in Kathmandu Valley, Nepal. *Environmental Earth Sciences 62*(8), 1723–1732.

Pandey, V. P., Chapagain, S. K. & Kazama, F. (2010). Evaluation of groundwater environment of Kathmandu Valley. *Environmental Earth Sciences 60*(6), 1329–1342.

Pandey, V. P., Shrestha, S. & Kazama, F. (2012). Groundwater in the Kathmandu Valley: development dynamics, consequences and prospects for sustainable management. *European Water 37*, 3–14.

Pathak, D. R., Hiratsuka, A. & Yamashiki, Y. A. (2011). Influence of anthropogenic activities and seasonal variation on groundwater quality of Kathmandu Valley using multivariate statistical analysis. In: *Conference Proceedings of Symposium H04 Held During IUGG2011*, July, Melbourne, Australia.

Schlumberger Water Services (2014). *Visual MODFLOW Flex – Integrated Conceptual & Numerical Groundwater Modeling.* Waterloo Hydrogeologic, Kitchener, Canada, p. 551.

Shrestha, S. D. (2012). Geology and hydrogeology of groundwater aquifer in Kathmandu valley. In: *Kathmandu Valley Groundwater Outlook.* Pandey, V. P., Shrestha, S. & Pradhananga, D. (eds). Asian Institute of Technology (AIT), The Small Earth Nepal (SEN), Center of Research for Environment Energy and Water (CREEW), International Center for River Basin Environment-University of Yamanashi (ICRE-UY), Kathmandu, Nepal, pp. 21–30.

Shrestha, S. M., Rijal, K. & Pokhrei, M. R. (2015). Assessment of arsenic contamination in deep groundwater resources of the Kathmandu Valley, Nepal. *Journal of Geoscience and Environment Protection 3*, 79–89.

Shrestha, S., Semkuyu, D. J. & Pandey, V. P. (2016). Assessment of groundwater vulnerability and risk to pollution in Kathmandu Valley, Nepal. *Science of the Total Environment 556*, 23–35.

Shrestha, P. K., Shakya, N. M., Pandey, V. P., Birkinshaw, S. J. & Shrestha, S. (2017). Model-based estimation of land subsidence in Kathmandu Valley, Nepal. *Geomatics, Natural Hazards Risk 5705*, 1–23.

Tamrakar, P. R. & Shakya, S. C. (2013). Physico-chemical assessment of deep groundwater quality of various sites of Kathmandu Metropolitan City, Nepal. *Research Journal of Chemical Science 3*(8), 78–82.

Tanaka, Y., Kei, N., Nakamura, T., Chapagain, K., Inoue, D., Sei, K., Mori, K., Sakamoto, Y. & Kazama, F. (2012). Characterization of microbial communities distributed in the groundwater pumped from deep tube wells in the Kathmandu Valley of Nepal. *Journal of Water and Health 10*(1), 170–180.

Thapa, B. R., Ishidaira, H., Pandey, V. P. & Shakya, N. M. (2016a). Impact assessment of Gorkha earthquake 2015 on portable water supply in Kathmandu Valley: preliminary analysis. *Journal of Japan Society of Civil Engineers Ser. 72*(4), 61–66.

Thapa, B. R., Ishidaira, H., Bui, T. H. & Shakya, N. M. (2016b). Evaluation of water resources in mountainous region of Kathmandu Valley using high resolution satellite precipitation product. *Journal of Japan Society of Civil Engineers, Ser. G (Environmental Research) 72*(5), 27–33.

Thapa, B. R., Ishidaira, H., Pandey, V. P. & Shakya, N. M. (2017). A multi-model approach for analyzing water balance dynamics in Kathmandu Valley, Nepal. *Journal of Hydrology: Regional Studies 9*, 149–162.

Thapa, B. R., Ishidaira, H., Pandey, V. P., Bhandari, T. M. & Shakya, N. M. (2018). Evaluation of Water Security in Kathmandu Valley before and after water transfer from another basin. *Water 10*(2), 224. doi:10.3390/w10020224

Udmale, P., Ishidaira, H., Thapa, B. & Shakya, N. (2016). The status of domestic water demand: supply deficit in the Kathmandu Valley, Nepal. *Water 8*(5), 196.

Warner, N. R., Levy, J., Harpp, K. & Farruggia, F. (2008). Drinking water quality in Nepal's Kathmandu Valley: a survey and assessment of selected controlling site characteristics. *Hydrogeological Journal 16*(2), 321–334.

WHO (2005). Minimum water quantity needed for domestic uses. WHO Regional Office for South-East Asia. How much water is needed? WHO/SEARO Technical Notes for Emergencies no. 9, pp. 1–4.

Yoshida, M. & Igarashi, Y. (1984). Neogene to Quaternary lacustrine sediments in the Kathmandu Valley, Nepal. *Journal of Nepal Geological Society 7*, 73–100.

IWA Publishing's authorised EU representative for General Product Safety Regulations is Diane D'Arras, 15 rue Duret, 75116 Paris, France, e-mail: safety@iwap.co.uk.

Printed and bound by CPI Group (UK) Ltd, Croydon, CR0 4YY
12/05/2026
02108870-0020